Johannes Ofner

Secondary organic aerosol & atmospheric halogen species

Johannes Ofner

Secondary organic aerosol & atmospheric halogen species

Formation of secondary organic aerosol and its processing by atmospheric halogen species - A spectroscopic study

Südwestdeutscher Verlag für Hochschulschriften

Impressum/Imprint (nur für Deutschland/only for Germany)
Bibliografische Information der Deutschen Nationalbibliothek: Die Deutsche Nationalbibliothek verzeichnet diese Publikation in der Deutschen Nationalbibliografie; detaillierte bibliografische Daten sind im Internet über http://dnb.d-nb.de abrufbar.
Alle in diesem Buch genannten Marken und Produktnamen unterliegen warenzeichen-, marken- oder patentrechtlichem Schutz bzw. sind Warenzeichen oder eingetragene Warenzeichen der jeweiligen Inhaber. Die Wiedergabe von Marken, Produktnamen, Gebrauchsnamen, Handelsnamen, Warenbezeichnungen u.s.w. in diesem Werk berechtigt auch ohne besondere Kennzeichnung nicht zu der Annahme, dass solche Namen im Sinne der Warenzeichen- und Markenschutzgesetzgebung als frei zu betrachten wären und daher von jedermann benutzt werden dürften.

Verlag: Südwestdeutscher Verlag für Hochschulschriften GmbH & Co. KG
Heinrich-Böcking-Str. 6-8, 66121 Saarbrücken, Deutschland
Telefon +49 681 37 20 271-1, Telefax +49 681 37 20 271-0
Email: info@svh-verlag.de

Approved by: Bayreuth, University, PhD thesis, 2011

Herstellung in Deutschland:
Schaltungsdienst Lange o.H.G., Berlin
Books on Demand GmbH, Norderstedt
Reha GmbH, Saarbrücken
Amazon Distribution GmbH, Leipzig
ISBN: 978-3-8381-2992-1

Imprint (only for USA, GB)
Bibliographic information published by the Deutsche Nationalbibliothek: The Deutsche Nationalbibliothek lists this publication in the Deutsche Nationalbibliografie; detailed bibliographic data are available in the Internet at http://dnb.d-nb.de.
Any brand names and product names mentioned in this book are subject to trademark, brand or patent protection and are trademarks or registered trademarks of their respective holders. The use of brand names, product names, common names, trade names, product descriptions etc. even without a particular marking in this works is in no way to be construed to mean that such names may be regarded as unrestricted in respect of trademark and brand protection legislation and could thus be used by anyone.

Publisher: Südwestdeutscher Verlag für Hochschulschriften GmbH & Co. KG
Heinrich-Böcking-Str. 6-8, 66121 Saarbrücken, Germany
Phone +49 681 37 20 271-1, Fax +49 681 37 20 271-0
Email: info@svh-verlag.de

Printed in the U.S.A.
Printed in the U.K. by (see last page)
ISBN: 978-3-8381-2992-1

Copyright © 2011 by the author and Südwestdeutscher Verlag für Hochschulschriften GmbH & Co. KG and licensors
All rights reserved. Saarbrücken 2011

*"Ich bin immer noch verwirrt,
aber auf einem höheren Niveau."*
ENRICO FERMI (1901–1954)

Acknowledgments

I would like to thank Prof. Dr. Cornelius Zetzsch for the opportunity to work at the Atmospheric Chemistry Research Laboratory of the University of Bayreuth and for supervising my thesis.

Further, I would like to express my thanks to Prof. Dr. Hinrich Grothe, Institute for Materials Chemistry of the Vienna University of Technology, for his support in interpreting the infrared spectra and for very fruitful discussions.

I especially appreciate the discussions with Prof. Dr. Andreas Held, extending my knowledge of secondary organic aerosol formation.

My special thanks to Heinz-Ulrich Krüger for his support during the smog chamber runs by operating the CNC-DMPS system, and for very useful discussions and assistance as well as for being open for any idea I had.

I am also indebted to Prof. Dr. Heinz Friedrich Schöler and all other members of the HALOPROC project for fruitful discussions of my results and suggestions on further research.

Special thanks to Dr. Ph. Schmitt-Kopplin, Helmholtz Centrum Munich, for measuring the ultra-high-resolution mass spectra as well as for helping to interpret them. Thanks to Karin Whitmore, USTEM (Vienna University of Technology), for acquiring the FEG-SEM images.

I also have to thank Natalja Balzer and Joelle Buxmann for the SOA-halogen interaction experiments in the Teflon chamber. I want to thank Gerhard Küfner for his support. Further I wish to thank all my colleagues for their support and my friends in Germany and Austria for their encouragement.

Very special thanks to my mother Irene Ofner and my girlfriend Katharina Westermayer for their unconditional support, encouragement, and their understanding for me working for my Ph.D. thesis in Germany.

Abstract

Atmospheric aerosols play an important role in the global climate system. Through their physicochemical properties, they contribute in various ways to climate change and radiative forcing. Those properties can be considerably changed by processing the aerosols, which is especially significant for organic aerosols processed with atmospheric trace gases like halogens released through sea-salt activation or from other sources.

Based on aerosol smog-chamber experiments, the formation of secondary organic aerosols (SOA) from predominantly aliphatic (α-pinene) or aromatic (catechol and guaiacol) precursors and the processing of those model SOAs with simulated molecular and naturally released halogens were studied.

Different physicochemical methods were used to study the transformation of those organic aerosols. Infrared and UV/VIS spectroscopy allowed the determination of functional and structural changes during aerosol formation and processing. Using electron microscopy and measurement of the particle size distribution, the formation of the morphology and geometry of the particles was investigated. Temperature-programmed pyrolysis mass spectroscopy and ultra-high-resolution mass spectroscopy delivered detailed information on functional groups, extent of halogenation, and the macromolecular structure.

Organic aerosols are significantly influenced by atmospheric halogens. Halogen species from different sources change the aerosol size distribution, the presence of functional groups, and the optical properties. Furthermore, they even form halogenated species in the solid phase of the organic aerosol.

Zusammenfassung

Atmosphärische Aerosole spielen eine bedeutende Rolle im Klimasystem. Sie tragen auf Grund ihrer physikalisch-chemischen Eigenschaften in unterschiedlicher Weise zum Klimawandel und zum Strahlungshaushalt bei. Diese Eigenschaften ändern sich, wenn Aerosole (besonders organische Aerosole) mit Gasphasen-Spezies (z.B. Halogene, die aus Seesalzaerosol oder anderen Quellen freigesetzt werden) reagieren.

Basierend auf Aerosolkammerexperimenten wurden die Bildung und die Reaktionen von sekundären organischen Aerosolen (SOA) aus überwiegend aliphatischen (α-Pinen) oder aromatischen (Brenzkatechin und Guajakol) Vorläufersubstanzen mit natürlich freigesetzten molekularen Halogenen untersucht.

Verschiedene physikalisch-chemische Methoden wurden eingesetzt, um die Bildung und Umsetzung zu studieren. Infrarotspektroskopie und optische Spektroskopie ermöglichen die Untersuchung der funktionellen Gruppen und strukturellen Elemente während der Aerosolbildungs- und -reaktionsprozesse. Elektronenmikroskopie und die Messung der Aerosolgrößenverteilung liefern morphologische und geometrische Informationen über die Partikel. Temperaturprogrammierte Pyrolyse mit Massenspektrometrie sowie ultrahochauflösende Massenspektrometrie erlauben Aussagen über funktionelle Gruppen, den Zustand der Halogenierung sowie die makromolekulare Struktur.

Organische Aerosole werden durch atmosphärische Halogene stark beeinflusst. Halogene aus unterschiedlichen Quellen verändern die Größenverteilung, die funktionellen Gruppen und die optischen Eigenschaften. Halogenierte Substanzen werden dabei auch in der Partikelphase gebildet.

Contents

Acknowledgments   iii

Abstract   v

Contents   vii

1 Introduction   1
- 1.1 Atmospheric aerosols - general overview 1
- 1.2 Secondary organic aerosols (SOA) in the atmosphere 3
    - 1.2.1 SOA from terpene-type precursors 5
    - 1.2.2 SOA from other precursors 7
- 1.3 Atmospheric HULIS and corresponding models 9
    - 1.3.1 Organic acids and HULIS 11
- 1.4 Aging and chemical processing of organic aerosols 12
- 1.5 Atmospheric halogen chemistry 14
    - 1.5.1 Halogen release from sea-salt aerosol and salt pans 15
    - 1.5.2 Halogen interaction with atmospheric gases, aerosols, and organics 17
- 1.6 Scope of this work 19

2 Methods   21
- 2.1 Aerosol smog chambers 22
    - 2.1.1 700 L glass chamber 22
    - 2.1.2 3500 L Teflon chamber 25
- 2.2 Aerosol flow reactor 25
    - 2.2.1 Circular multi-reflection cell 27
- 2.3 Particle sizing 29
- 2.4 Electron microscopy of SOA 29
- 2.5 Fourier-transformation infrared spectroscopy (FTIR) 30
    - 2.5.1 Long-path infrared absorption spectroscopy in the 700 L glass chamber . 31
    - 2.5.2 Attenuated total reflectance (ATR) spectroscopy 31
    - 2.5.3 Electrostatic precipitators 32

2.6 Temperature-programmed pyrolysis mass spectroscopy 35
2.7 UV/VIS spectroscopy using an integrating sphere 37
2.8 Ultra-high-resolution mass spectroscopy 39
2.9 Sample preparation techniques 39

3 SOA formation 41
3.1 Experimental setup and materials 41
3.2 SOA from α-pinene 42
    3.2.1 Particle formation in the 700 L aerosol smog chamber 42
    3.2.2 Infrared spectroscopy of particle formation in the smog chamber 43
    3.2.3 Spectroscopy of particle formation using the aerosol flow reactor 45
    3.2.4 Spectroscopy of functional groups of particulate matter 47
    3.2.5 Optical properties 50
3.3 SOA from catechol and guaiacol 51
    3.3.1 Particle formation - particle number concentration and size distributions 51
    3.3.2 Particle imaging using FEG-SEM 53
    3.3.3 Infrared spectroscopy of particle formation in the smog chamber 54
    3.3.4 Spectroscopy of particle formation using the aerosol flow reactor 59
    3.3.5 Spectroscopy of functional groups of SOA particles 61
    3.3.6 Optical properties 66
    3.3.7 Ultra-high-resolution mass spectra of SOA from catechol and guaiacol . . 68

4 Molecular processing 71
4.1 Experimental setup 71
4.2 Influences on aerosol size and particle number distributions 72
4.3 Changes in the vibrational gas-phase and particle-phase spectra 73
4.4 Changes in the vibrational spectra of the particulate phase by reaction with halogens 77
4.5 Identification of halogenated compounds in the particle phase 79
4.6 Parameters of processed SOA calculated from ICR-FT/MS spectra 84
4.7 Changes in the aerosol optical properties 86

5 Simulated natural processing 89
5.1 Processing with halogens released from a simulated salt pan 89
    5.1.1 Experimental setup 89
    5.1.2 Aerosol size distribution 90
    5.1.3 ATR-FTIR spectroscopy 91
    5.1.4 Optical properties 92

	5.1.5 Mass spectroscopy	93
	5.1.6 Influence of organic aerosols on halogen release mechanisms	94
5.2	Processing with halogens released from simulated sea-salt aerosol	95
	5.2.1 Experimental setup	95
	5.2.2 Aerosol size distribution	96
	5.2.3 ATR-FTIR spectroscopy	96
	5.2.4 Optical properties	97
	5.2.5 Mass spectroscopy	98

6 Discussion and conclusions — **99**

6.1	Characterization of SOA models	99
	6.1.1 SOA formation and characterization	99
	6.1.2 HULIS-like behavior of SOA from catechol and guaiacol	101
6.2	Halogenation of SOA using molecular halogen species	105
6.3	Comparison of the results of natural halogenation processes to the molecular processing study	108
6.4	Influence of natural halogen species on organic aerosols	108
6.5	Applicability of aerosol smog-chamber studies and physicochemical methods	110

References — **113**

List of Figures — **133**

List of Tables — **139**

1 Introduction

1.1 Atmospheric aerosols - general overview

"Whereas an aerosol is technically defined as a suspension of fine solid or liquid particles in a gas, common usage refers to the aerosol as the particulate component only." (Seinfeld and Pandis, 2006)

While public discussion about atmospheric science was dominated by terms like "LA smog" and "ozone hole" for a long time, atmospheric aerosols attracted attention at last, when the Intergovernmental Panel on Climate Change released the 4th assessment report. Forster et al. (2007) summarized the influence of atmospheric aerosols on radiative forcing and concluded that a high uncertainty remains, while the earlier 3rd report (Ramaswamy et al., 2001) revealed an even lower comprehension of the impact of atmospheric aerosols on climate change and radiative forcing.

This is also indicated by the available textbooks on atmospheric science. Standard works like "Atmospheric Chemistry and Physics" by Seinfeld and Pandis (2006) and "Chemistry of the Upper and Lower Atmosphere" by Finlayson-Pitts and Pitts (2000) are mainly focused on homogeneous chemistry of the gas phase. Atmospheric aerosols are mentioned in the context of physical particle properties, like aerosol size distributions and particle dynamics, thermodynamics, chemical composition and nucleation. While some aspects of formation of secondary organic aerosols (SOA) are well known and widely published, heterogeneous reactions of gas-phase molecules with surfaces of the particulate matter of atmospheric aerosols are hardly reported.

This gap in detailed knowledge could be explained by the large diversity of interactions with the atmospheric environment those particles might take part in. Aerosols have direct effects on the climate system, e.g. through interaction with solar radiation or release of greenhouse-gases during their formation process. Also indirect effects are described, e.g. their ability to act as cloud condensation nuclei (CCN) or ice nuclei (IN) as well as absorbers of atmospheric trace gases and water vapor (Seinfeld and Pandis, 2006).

1 Introduction

Those interactions are quite variable, depending on the kind of aerosol. The chemical composition varies from primary dust or salt particles to heterogeneously formed inorganic molecular clusters and macromolecular organic particles with complex structure (see e.g. Andreae and Crutzen (1997), Pöschl (2005), Schnelle-Kreis et al. (2007)). Based on their source and chemical composition, aerosols are characterized as primary/secondary organic/inorganic aerosols (P/S O/I Aerosol). One major primary aerosol is the marine aerosol, consisting of sea-salt particles ejected by sea spray. The class of secondary particles is dominated by organic aerosols formed from terpenes and anthropogenic emissions (Schnelle-Kreis et al., 2007).

The diversity of interactions and the vast variety of aerosols require a large pool of analytical methods and experimental setups, as summarized in the literature (e.g. Nießner (1991), Finlayson-Pitts and Pitts (2000), Zellner et al. (2009)). Experimental setups range from Knudsen cells to flow reactors and aerosol smog chambers. Those different laboratory setups are described in detail by Grothe (2010). While the basic physicochemical experiments, using a well defined setup, give access to single kinetic or thermodynamical parameters, aerosol smog chambers exhibit the most realistic but also most complex setup for studying aerosol chemistry (Brauers and Wiesen, 2007).

The interaction of atmospheric aerosols with atmospheric trace gases or industrial process gases has only been studied by few research groups. Primary organic aerosols were examined with respect to automotive exhaust emissions to develop post-processing methods for soot particles (Ofner, 2006; Ofner and Grothe, 2007; Muckenhuber and Grothe, 2004, 2006, 2007). SOA formation and post-processing influenced by NO_x species was studied by Yu et al. (2008). George and Abbatt (2010) reviewed the heterogeneous processing of organic and inorganic aerosols with hydroxyl radicals. A detailed review by Finlayson-Pitts (2009) of some known surface reactions, also explaining the chemistry at the sea salt aerosol interface, is only one example of the available literature.

The large field of atmospheric aerosols with its various implications on the atmospheric environment and climate needs a large variety of experimental methods for the characterization of chemical formation and transformation processes. While some aspects of atmospheric aerosol, like their formation processes, are common research topics, others, like the heterogeneous processing of aerosol particles with atmospheric trace gases, were hardly looked at up to now. This thesis summarizes the investigation of the processing of some well-defined organic model aerosols with halogen species which are released from sea-salt aerosol or brine. Several important aerosol characteristics, like the evolution and transformation of the aerosol size distribution, the morphology, the functional and structural changes, the optical properties and the chemical composition, were monitored and discussed in detail. The following sections will introduce organic model aerosols (section 1.2) and halogen release mechanisms (section 1.5).

The experimental setups and methods used as well as the development of new methods for aerosol research are described in chapter 2. Formation procedures and characterization of the organic model aerosols can be found in chapter 3. To study single reaction steps and the influence of halogens on the organic aerosol, molecular processing of those aerosols with chlorine and bromine was investigated (see chapter 4). The results allowed to interpret the interaction of organic aerosols with halogens released from simulated natural sources (Chapter 5).

1.2 Secondary organic aerosols (SOA) in the atmosphere

Organic aerosols contribute significantly to the overall aerosol budget. Primary organic aerosol emissions from biomass burning, combustion engines, and even biogenic sources play an important role. Lary et al. (1999) summarized chemical functional features of carbonaceous aerosols, like soot, and their surface properties. However, this work is focused on aliphatic or aromatic SOA where particle formation takes place in the free troposphere from oxidized volatile organics.

Andreae and Crutzen (1997) estimated annual mass fluxes of 30–270 Tg y^{-1} by tropospheric oxidation of biogenic and anthropogenic volatile organic compounds. Contributions to the fine fraction of organic aerosol between 20–50 % at mid-latitudes and up to 90 % in tropical forested areas are reported by Kanakidou et al. (2005). Biogenic secondary organic carbon (BSOC, e.g. from biogenic terpenes) fluxes are estimated to be approximately 88 Tg C y^{-1} (Hallquist et al., 2009). α-Pinene contributes up to 58 % to the overall monoterpene budget (Kanakidou et al., 2005). With respect to aromatic SOA precursors, Kanakidou et al. (2005) report anthropogenic emissions of 6.7 Tg y^{-1} toluene, 4.5 Tg y^{-1} xylene, 0.8 Tg y^{-1} trimethylbenzene, and 3.8 Tg y^{-1} of other aromatic compounds. Total biogenic SOA fluxes of 12–70 Tg y^{-1} from bottom-up estimates are reported by Hallquist et al. (2009). Fluxes of 17 Tg C y^{-1} for SOA from biomass burning and 10 Tg C y^{-1} for SOA from anthropogenic sources are reported based on a top-down approach (Hallquist et al., 2009).

The formation of aerosol from volatile organic precursors by oxidation is the established nucleation process (Seinfeld and Pandis, 2006). The term homogeneous nucleation denotes particle formation from supersaturated vapors without seed particles. The nucleation process of an organic aerosol is a very complex mechanism, implying multiple condensation, reaction, and desorption steps of the organic precursor and its oxidized derivative (Kroll and Seinfeld, 2008). The complex transformations are shown in figure 1.1, which has been adapted from Kroll and Seinfeld (2008). The multiple oxidation steps in the gas and particle phase lead to compounds of very low volatility. The overall oxidation ends up in simple carbon-containing molecules like CO_2 or CO.

1 Introduction

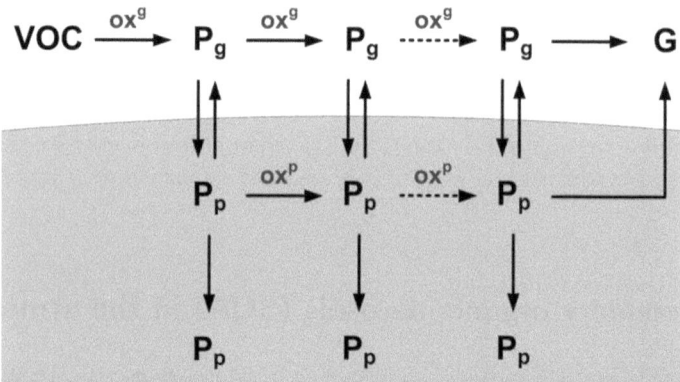

Figure 1.1: Current understanding of the oxidation steps (gas phase: $ox.^g$; particle phase: $ox.^p$) starting from a SOA precursor (VOC) leading to low-volatile compounds in the gas phase (P_g), partitioning to the particle phase (P_p) and finally releasing simple gaseous molecules (G) like CO_2 or CO. Adapted from Kroll and Seinfeld (2008).

Apart from the oxidation of the precursor and the resulting intermediates, gas-particle partitioning takes place (Hallquist et al., 2009). While gas-particle partitioning models assume a liquid state of the SOA particles (Hallquist et al., 2009; Jimenez et al., 2009), latest research revealed the possibility of an amorphous or solid state at ambient conditions (Virtanen et al., 2010). The numerous interactions between gas molecules and emerging particles in the course of the nucleations makes it necessary to use the technical definition of an aerosol (see Seinfeld and Pandis (2006)), especially for aerosol smog-chamber experiments, where the gas phase cannot be separated from the particles.

Beside their complex formation process, organic aerosols exhibit the formation of oligomeric structures as well as photochemical aging and reactivity towards atmospheric trace gases. Based on their Teflon chamber experiments Gao et al. (2004b) concluded, that oligomers might be widely present in atmospheric organic aerosol. Oligomers seem to be formed through acid- or base-catalyzed heterogeneous reactions. Heaton et al. (2007) found dimers and higher-molecular-weight oligomers in laboratory-generated SOA from monoterpenes, and linked them to atmospheric humic-like substances (HULIS) (see section 1.3). Kalberer et al. (2004) report that 50 % of SOA from aromatic oxidation is composed of polymers. Two precursor types, the terpenes and the aromatics, seem to form oligomers and thus be related to HULIS.

1.2.1 SOA from terpene-type precursors

A large variety of different biogenic emitted terpenes are reported (Seinfeld and Pandis, 2006; Steinbrecher and Koppmann, 2007). All these terpenes are based on the structural element of isoprene, forming e.g. α- and β-pinene, Δ^3-carene, sabinene, limonene and p-cymene.

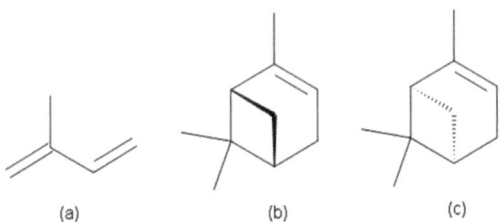

Figure 1.2: The terpene precursor isoprene (a) and the monoterpenes (-)-α-pinene (b) and (+)-α-pinene (c). The racemate of those two structural isomers, called α-pinene, is widely used as model compound in laboratory experiments for SOA research.

As mentioned above, a very important monoterpene is α-pinene (see figure 1.2). This bicyclic terpenoid has been characterized in detail, e.g. the vibrational spectrum is well known and nearly completely unraveled (Wilson, 1976). The reactivity towards oxidizing atmospheric trace gases has been studied in several aerosol smog-chamber studies (e.g. Nolting et al. (1988)). A suggestion for the rate constant of the reaction with ozone of 8.52×10^{-17} cm^3 s^{-1} (Atkinson, 1994) at room temperature has been based on values of 8.2×10^{-17} (Nolting et al., 1988) and 9.71×10^{-17} cm^3 s^{-1} (Atkinson, 1994). Since OH radicals are released in the early steps of the reaction even in the absence of NO_x (Nolting and Zetzsch, 1989), OH yields from α-pinene ozonolysis ranging from 68 % (Berndt et al., 2003) to 91 % (Siese et al., 2001), independent of the presence of H_2O, have been reported for this process. The transformation pathways of this model substance have been well characterized by many authors, and various gaseous intermediates, such as pinonaldehyde, pinene oxide, norpinaldehyde and 4-oxopinonaldehyde, have been summarized in a recent overview (Yu et al., 2008), although the reported yields of pinonaldehyde span a range of 3–53 % in the O_3-reaction and 6–87 % in the OH reaction.

Based on the gas-phase kinetics of the reaction of the monoterpenoid precursor with atmospheric oxidizing trace gases, the potential and yields of particle formation were studied using stopped-flow systems or aerosol smog chambers (e.g. Berndt et al. (2003); Yu et al. (2008)). Jonsson et al. (2007) studied the change of volatility during the aerosol formation process. Time-resolved infrared spectroscopy was used by Sax et al. (2005) to study the change of chemical functionality during and after the aerosol formation process. They report that the chemical composition of the aerosol continues to change even after the particle formation process has

1 Introduction

finished. Further, a detailed analysis of the resulting SOA particle infrared spectrum was performed by Sax et al. (2005), identifying several oxygen-containing functional groups. Based on these studies, a detailed picture of the α-pinene oxidation process is presented by Berndt et al. (2003) and Yu et al. (2008). It is widely accepted that the ozone oxidation of a double bond forms a primary ozonide which rearranges to an exited Criegee intermediate (Atkinson, 1997). This Criegee intermediate replaces the carbon-carbon double bond by an ether- and a peroxo-group. However, the formation of this intermediate seems to be less important for the α-pinene oxidation and SOA formation (Berndt et al., 2003).

When α-pinene was established as a useful precursor for laboratory studies on SOA formation and processing, this model aerosol was used to investigate its properties and its reactivity towards the atmospheric environment. Iinuma et al. (2009) studied the formation of organosulfates, using oxidized species of α-pinene on the surface of a wet seed aerosol. The temperature- and humidity-dependent aerosol formation from this monoterpene precursor was investigated by Jonsson et al. (2008). Juranyi et al. (2009) underlined the importance of gas-to-particle partitioning of SOA from α-pinene on the hygroscopicity and the droplet activation behavior of this aerosol. As already mentioned, the investigation of the oligomeric compound formation is very important in SOA research and thus, α-pinene SOA was used to study the formation processes. Oligomers with a molecular weight above 250 Daltons were found by Gao et al. (2004a) and were associated with seed aerosol acidity. The importance of particle acidity for the formation oligomeric compounds was underlined by Gao et al. (2004b) and Iinuma et al. (2004).

Other terpenes, like limonene, Δ^3-carene and even isoprene were also used to characterize SOA formation pathways, oligomer formation, temperature dependence, and multiphase chemistry. Pan et al. (2009) demonstrated the reactivity of organic peroxides with $+RO_2$ groups or HO_2 in SOA from limonene precursor as the dominant route to photochemically active species. Maksymiuk et al. (2009) describe ring-opening reactions and the formation of a Criegee intermediate for the oxidation of limonene. The sabinene-OH reaction was studied by Carrasco et al. (2006). Δ^3-carene was used in addition to α-pinene to study e.g. oligomer formation (Heaton et al., 2007) or temperature- and humidity-dependence of the SOA formation (Jonsson et al., 2008). Latest research focused on isoprene, the most simple terpene. In contrast to Pandis et al. (1991), Claeys et al. (2004) found a strong contribution of isoprene to SOA in the Amazonian rain forest. Kiendler-Scharr et al. (2009) report an inhibition of new particle formation by the high reactivity of isoprene with OH radicals. Paulot et al. (2009) report epoxide formation during the photo-oxidation of isoprene.

Based on the amount of available studies and the widespread acceptance of this model, α-pinene was chosen as precursor for a terpenoid-based SOA in the present study. As mentioned

1.2 Secondary organic aerosols (SOA) in the atmosphere

below in detail, SOA from α-pinene was formed by homogeneous nucleation, aged to achieve a stable chemical state, and characterized with regard to its physicochemical properties relevant for this study.

1.2.2 SOA from other precursors

Apart from terpenes, especially aromatic and olefinic compounds became very important as precursors to study the formation of SOA. Many investigations focus on the atmospheric oxidation of aromatic compounds (e.g. Becker and Klein (1987); Lay and Bozzelli (1996)). An overview of the tropospheric oxidation of aromatic compounds is given by Seinfeld and Pandis (2006). McDow et al. (1994) analyzed the decomposition of polycyclic aromatic hydrocarbons and located them as coatings on combustion-derived atmospheric particles. Forstner et al. (1997) studied the formation of SOA from toluene- and xylene-type precursors and discussed ring-opening mechanisms based on reactions with hydroxyl radicals. Johnson et al. (2005) simulated the formation of SOA from the photo-oxidation of aromatic hydrocarbons and concluded that there still is a lack of knowledge regarding the mechanisms underlying the formation of SOA in these systems. Benzene, p-xylene, and 1,3,5-trimethylbenzene were used to simulate SOA formation by the Master Chemical Mechanism (MCM) (Johnson et al., 2005). These results were compared to aerosol smog-chamber runs at the EUPHORE aerosol smog chamber in Valencia/Spain. Experiments with SOA from 1,3,5-trimethylbenzene indicated oligomer formation with high molecular weights (up to 1000 Da) (Baltensperger et al., 2005). SOA formation from m-xylene, toluene, and benzene under heterogeneous nucleation conditions on seed particles with varying acidity was studied by Ng et al. (2007). They state that a detailed analysis of the chemical composition of the aromatic SOA is important for understanding the formation pathways. Studies on formation and processing of those precursors with respect to SOA formation in the presence of NO_x have been performed by Jang and Kamens (2001) and Ng et al. (2007). Even mixtures of terpenes and aromatic hydrocarbons were used to study SOA formation. Seinfeld et al. (2003) used a mixture of α-pinene and toluene in a continuous-flow chamber. Offenberg et al. (2007) used the same mixture to evaluate an organic tracer method. Formation of airborne polymers from photo-oxidation of aromatic precursors was reported by Kalberer et al. (2004). SOA formation pathways from aromatic hydrocarbons like benzene, xylene, and toluene were added to a global chemical transport model by Henze et al. (2008). They estimated benzene to be the most important species for global aromatic SOA formation. The most important benzene-type SOA precursors investigated in literature are summarized in figure 1.3.

Apart from purely aromatic systems, also cycloalkenes were used with and without admixture of terpenes to study SOA formation by Gao et al. (2004a,b), who found an indication for

1 Introduction

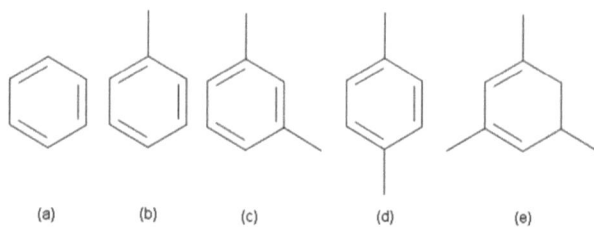

Figure 1.3: Some benzene-type precursors and selected literature on SOA formation: (a) benzene by Johnson et al. (2005), Baltensperger et al. (2005), Henze et al. (2008), and Ng et al. (2007); (b) toluene by Forstner et al. (1997), Henze et al. (2008), Ng et al. (2007), Seinfeld et al. (2003), and Jang and Kamens (2001); (c) m-xylene and (d) p-xylene by Ng et al. (2007), Forstner et al. (1997), Johnson et al. (2005), and Henze et al. (2008); (e) 1,3,5-trimethylbenzene by Johnson et al. (2005) and Baltensperger et al. (2005)

oligomer formation. Gao et al. (2004b) assumed an acid- or base-catalyzed heterogeneous reaction forming those oligomers from the terpene-alkene mixtures.

Several other precursors and model aerosol surfaces were used to investigate aerosol formation or heterogeneous processing. Vesna et al. (2009) used oleic acid aerosols to study the heterogeneous reaction with ozone in an aerosol flow reactor. Other organic acids like maleic and fumaric acid aerosol particles were also used (Najera et al., 2009). The conversion of 1,4-hydroxycarbonyls to hemiacetals and dihydrofurans on organic aerosol particles was investigated by Lim and Ziemann (2009).

Oxidized benzenes, like phenols, offer another class of SOA precursors. Olariu et al. (2000) measured the rate coefficients of the gas-phase reaction of OH with dihydroxybenzenes. Tomas et al. (2003) investigated the reaction of ozone with benzenediols and state that those secondary oxidation products might be important for SOA observed in urban areas. High oxygen-to-carbon ratios and high yields of oxalate are reported for SOA formation from phenolic precursors (Sun et al., 2010). They report dimer formation in aqueous-phase reactions of different phenols, including guaiacol (1-hydroxy,2-methoxy-benzene). Gas-phase chemistry and SOA formation from nitrophenols and catechol (1,2-dihydroxybenzene) was studied by Bejan (2006). Recent studies on aerosol formation from catechol (Coeur-Tourneur et al., 2009) obtained high mass yields ranging from 17–86 % in a smog chamber in the presence of ozone with only minor influence of self-produced OH, which is known from scavenger experiments to enhance the consumption of catechol by 30 % (Tomas et al., 2003). Aerosol formation yields from catechol of 20–58 % have been obtained in experimental runs at EUPHORE (Olariu et al., 2004). Also first-order products were identified during this campaign like formaldehyde, acetaldehyde, glyoxal, or muconic acid. The rate constant for the reaction of catechol with O_3 has been determined to be 9.6×10^{-18} cm^3 s^{-1} at 298 K, and the vicinal OH groups have been suggested as a potential cause of the high reactivity of catechol towards ozone (Tomas et al., 2003). Furthermore, infrared spectroscopy of the solid phase oxidation products of catechol has been performed in the aqueous phase (Khovratovich et al., 1998). The main phenol-type precursors mentioned above are shown in figure 1.4.

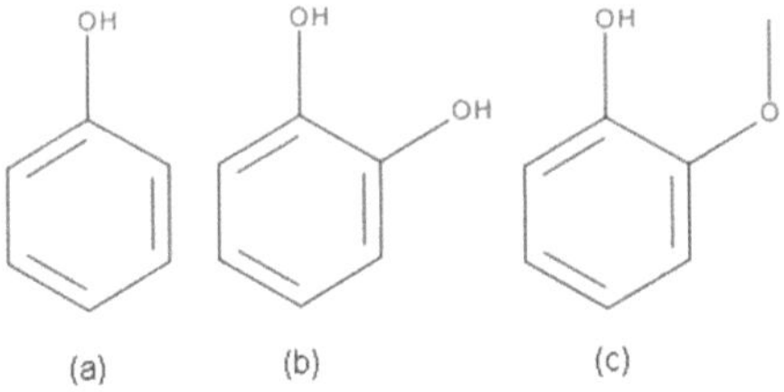

Figure 1.4: Some phenol-type precursors: (a) phenol; (b) catechol; (c) guaiacol

1.3 Atmospheric HULIS and corresponding models

HULIS represent a special class of organic aerosols, including macromolecular organic substances in atmosphere and rainwater (Mukai and Ambe, 1986). The current state-of-the-art model of atmospheric HULIS is based on selected macromolecular structures with an aromatic or olefinic core. These substances were found in water extracts of filter samples and exhibit an aromatic macromolecular structure of the organic matter. Because of their similarity to humic substances in soil, these aerosols were called HULIS (humic-like substances). Mukai and Ambe (1986) suggested a polycyclic structure with hydrocarbon side chains and hydroxyl-, carbonyl-, and hydroxy-containing functional groups. While atmospheric HULIS seem to be formed by rapid abiotic oxidation pathways in the atmosphere (Graber and Rudich, 2006), appropriate aromatic or olefinic precursors are needed to describe their formation pathways. Also Kroll and Seinfeld (2008) concluded that the high aromatic content of HULIS might originate from oxidative and non-oxidative particle-phase reactions of different precursors including aromatics.

While terpenes are a sizable source of unsaturated aliphatic compounds, a possible source of aromatic precursors might be open biomass burning, which was recognized as major primary source (Salma et al., 2010) - the so called biomass-burning organic aerosol (BBOA). Another prominent component resulting from biomass-burning events as reported by these authors are secondary HULIS, formed by photochemical reactions of volatile organic compounds (VOC) followed by oligomerization in the gas phase or by heterogeneous processing on seed aerosol

1 Introduction

particles. A detailed overview of different sources and formation pathways and corresponding studies is given by Salma et al. (2010). Stone et al. (2009) report emissions from motor vehicles, biomass burning, and SOA as important sources of HULIS precursors.

Since the research topic of HULIS first attracted attention, that special class of organic aerosols has been characterized in detail. Samburova et al. (2007) investigated the functional groups of HULIS and concluded that carboxylic, arylic, phenolic, and aliphatic groups contribute up to 14 % to the total mass of HULIS. Duarte et al. (2005) characterized the water-soluble organic matter isolated from atmospheric aerosols using Fourier-transformation infrared (FTIR) spectroscopy. Havers et al. (1998) described the IR spectrum of HULIS on airborne dust in detail and recorded UV/VIS spectra, finding absorption even in the range of visible light. Different oxidized aromatic compounds, like benzoic acid, methoxy-benzoic acid, benzenedicarboxylic acids, and benzenetricarboxylic acids, were identified in HULIS samples by Gelencser et al. (2000), using pyrolysis-gas-chromatography mass spectroscopy. Baduel et al. (2009) report $\pi - \pi^*$ transitions of aromatic compounds in the UV/VIS spectra of HULIS.

Humic and fulvic acids from soil, like Suwannee River fulvic acid (SRFA), have been the state-of-the-art model compounds for atmospheric HULIS in laboratory experiments (see e.g. Chan and Chan (2003) and Dinar et al. (2006)). But they seem to differ significantly from atmospheric HULIS, as reported by a detailed and critical review about the humic-like character of atmospheric HULIS (Graber and Rudich, 2006). The main differences in aromaticity, oxidation state, and macromolecular structure are caused by the fast heterogeneous formation pathway of HULIS by oxidation in contrast to the formation pathway of humic substances in soils. Thus, precursors are needed as in-situ model substances to generate SOA with HULIS qualities (Cowen and Al-Abadleh, 2009).

Gelencser et al. (2003) report the formation of light-absorbing organic matter from aromatic hydroxy acids with hydroxyl radicals and propose the products to be HULIS. Hoffer et al. (2004) characterized the Fenton-reaction products of 3,5-dihydroxybenzoic acid with OH radicals as synthetic HULIS. Baltensperger et al. (2005) studied the SOA formation from the photo-oxidation of an anthropogenic (1,3,5-trimethylbenzene) and a biogenic (α-pinene) precursor and concluded that oligomer formation without acidic seed particles is one of the major sources of atmospheric HULIS. Cowen and Al-Abadleh (2009) characterized photodegradated tannic acid as a model for HULIS.

All these precursors with their respective molecular structures represent state-of-the-art models featuring an aromatic core, aliphatic side chains, and oxygen-containing functional groups. To generate an organic aerosol which allows to study the physicochemical properties and reactions of the aromatic core without any aliphatic side chains, other model compounds are needed.

1.3 Atmospheric HULIS and corresponding models

Various aromatic compounds are reported in fine particle emissions from fireplace combustion by Fine et al. (2002). In their detailed table of organic compounds, guaiacol and substituted guaiacols as well as catechol and other hydroxybenzenes can be found. The same compounds were discussed by Hays et al. (2005) and Fine et al. (2004). Hoffer et al. (2006) studied the optical properties of HULIS derived from BBOA and concluded that those HULIS could play an active role in the radiative transfer and in photochemical processes. Thus, catechol and its methyl-ether derivative guaiacol (both were introduced in section 1.2.2 and illustrated in figure 1.4) could be appropriate candidates for HULIS precursors. While catechol is used as a model compound for soils (Huber et al., 2010), atmospheric oxidation and ring-opening processes could transform it into aerosols with olefinic structure, which are known from BBOA. The transformation of catechol into humic polymers in the context of humus formation was studied by Ahn et al. (2006). The aerosol formation from catechol has been studied very recently (Coeur-Tourneur et al., 2009). Nieto-Gligorovski et al. (2008, 2010) studied oxidation reactions of 4-carboxybenzophenone/catechol films using UV/VIS and FTIR spectroscopy. They report a photo-sensitized oxidation of the phenolic precursor by ozone in the presence of simulated sunlight forming products with properties similar to HULIS.

Consequently, catechol and guaiacol were chosen as precursors to study heterogeneous reactions with the aromatic or olefinic core of atmospheric HULIS. The aerosol originating from those two precursors was characterized in detail using different methods described in chapter 2.

1.3.1 Organic acids and HULIS

Carboxylic groups are of special interest because of their influence on e.g. water activity and surface tension (Salma and Lang, 2008). The authors highlighted the importance of organic acids in atmospheric aerosols, especially in HULIS. Limbeck et al. (2005) report the gas-to-particle distribution of low-molecular dicarboxylic acids with semi-volatile behavior at two different sites in Europe. Coury and Dillner (2009) found out that aldehydes/ketones, carboxylic acids, esters/lactones, and acid anhydrides account for up to 20 % of the organic mass in their aersol samples. Fisseha et al. (2004) used 1,3,5-trimethylbenzene as SOA precursor and determined a contribution of 20–45 % of organic acids to the overall aerosol mass. In their HULIS samples, Stone et al. (2009) distinguished between aromatic and aliphatic carboxylic acids and allocated them to different sources. Limbeck and Puxbaum (1999) found mono- and dicarboxylic acids at three different sampling sites. The semi-volatile behaviour of low-molecular-weight dicarboxylic acids is important for their partitioning between gas and aerosol phase (Limbeck et al., 2005). The importance of dicarboxylic acids, ketocarboxylic acids, and dicarbonyls in the range of C_2-C_{11} was demonstrated by Kundu et al. (2010). Kumagai et al.

1 Introduction

(2010) found high values of dicarboxylic acids in fine particles in the Kanto plain region (Japan) and correlated them with the secondary formation of water-soluble organic carbon (WSOC) in the atmosphere. Also, hydroxydicarboxylic acids were established as markers for SOA (Claeys et al., 2007).

Thus, carboxylic acids are strongly correlated to secondary formation pathways of atmospheric HULIS, and analyzing aromatic and aliphatic carboxylic groups is very important to characterize the chemical state of the aerosol.

1.4 Aging and chemical processing of organic aerosols

The physicochemical transformation of SOA precursors or SOA aerosol particles is not finished with particle formation. Chemical aging and chemical post-processing takes place during the atmospheric lifetime of the aerosol. Baltensperger et al. (2005) found increasing thermal stability of SOA samples from α-pinene and 1,3,5-trimethylbenzene during photochemical aging. A scheme of SOA formation pathways, set up by Donahue et al. (2006), takes chemical aging in the condensed phase into account, underlining the contribution of oxidized "intermediate" VOC (IVOC) to the overall SOA formation process. The contribution of photochemically aged IVOC and "semi-volatiles" (SVOC) to SOA, and the resulting excess of traditionally calculated SOA formation yields was discussed by Robinson et al. (2007). Andreae (2009) states that aging of organic aerosols from different sources eventually leads to remarkably similar properties.

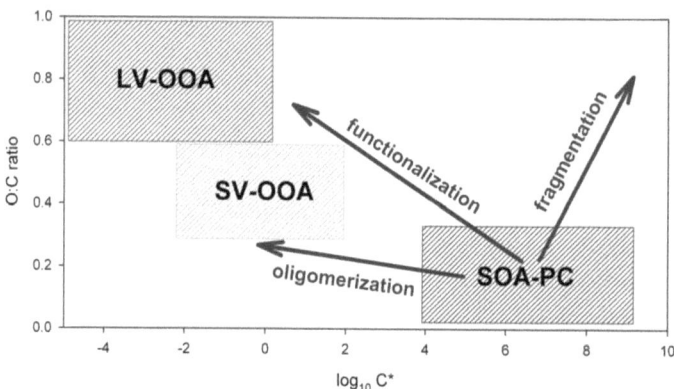

Figure 1.5: 2-D plot of aerosol formation and aging; change of O/C ratio and saturation concentration C^* of the aerosol precursor (SOA-PC) by functionalization and oligomerization; formation of SV- and LV-OOA in the related O/C-C^* space. Adapted from Jimenez et al. (2009).

1.4 Aging and chemical processing of organic aerosols

An attempt at a general road map of photochemical aging was given by Jimenez et al. (2009), when they correlated the proceeding of aging of the SOA precursor with an increasing O/C ratio and a decreasing saturation concentration (Figure 1.5). Their 2-D framework describes the aging of first-order generation products to SV-OOA (semi-volatile oxidized organic aerosol; O/C: 0.3 to 0.6; saturation concentration: 10^2 to 10^{-1} μg m^{-3}). The second aging step to LV-OOA (low-volatile oxidized organic aerosol; O/C: 0.6 to 1; saturation concentration: 10^{-2} to 10^{-5} μg m^{-3}) is described by condensed-phase OH reactions. Ng et al. (2010) used factor analysis of aerosol mass spectra to follow the aging of organic aerosols in the atmosphere. Factor f_{44} represents the ion 44 (CO_2^+) and factor f_{43} the ion 43 ($C_2H_3O^+$). LV-OOA exhibits high values of factor f_{44} and low values of factor f_{43}. Recently, Kroll et al. (2011) proposed to use the average carbon oxidation state ($\overline{OS_C}$) to describe the oxidation of SOA precursors and aging of SOA particles (see figure 1.6). The average carbon oxidation state is better suited to characterize chemical changes resulting from oxidation than O/C ratios or factor analysis. According to equation 1.1, the carbon oxidation state is defined as the negative sum of the molecular fraction of every atomic species n_i related to the number of carbon atoms n_C multiplied with the state of oxidation OS_i of component i, where j is the number of every atomic species, at their different oxidation states except carbon. For molecules containing only carbon, oxygen, and hydrogen, the calculation can take a simplified form (equation 1.2).

$$\overline{OS_C} = -\sum_{i=1}^{j} OS_i \frac{n_i}{n_C} \qquad (1.1)$$

$$\overline{OS_C} = 2\frac{n_O}{n_C} - \frac{n_H}{n_C} \qquad (1.2)$$

Beside aging of organic aerosols, chemical processing with other trace gases must be taken into account. The heterogeneous reaction of organic aerosols with nitrogen species might be defined as aging. Bröske et al. (2003) studied the heterogeneous reactions of catechol-derived aerosol with nitrogenous trace gases, especially the heterogeneous conversion of NO_2 in the presence of ozone. Rudich (2003) summarized reactive uptake experiments on organic aerosols using O_3, OH, NO_2, $ClONO_2$, NO_3, and also Cl and Br as reactants. Ishikawa et al. (1986) used different derivatives of benzene, including catechol, to study the formation of chloroacetic acids. A detailed review on studies about heterogeneous reactions of organics with halogen species is given in section 1.5.2.

1 Introduction

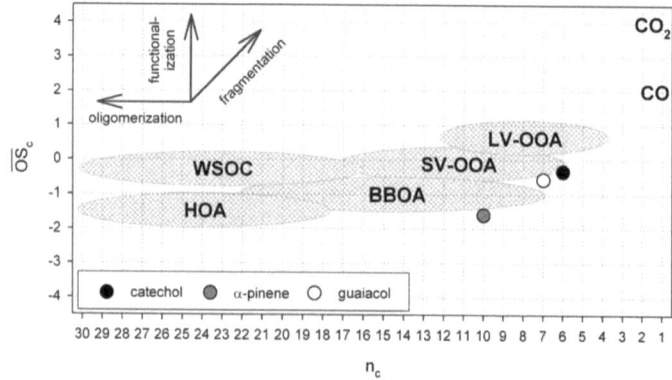

Figure 1.6: Different organic aerosol types (LV-OOA, SV-OOA, BBOA, WSOC, HOA (hydrocarbon-like organic aerosol)) and SOA precursors used in this study (catechol, guaiacol, α-pinene) in the average carbon oxidation state ($\overline{OS_C}$) and number of carbon atoms (n_C) space. Adapted from Kroll et al. (2011).

1.5 Atmospheric halogen chemistry

Due to their capability for stratospheric ozone depletion, reactive halogen species became a very important topic of atmospheric research. Starting with the first publications on ozone destruction catalyzed by chlorine atoms(Molina and Rowland, 1974; Rowland and Molina, 1975), this field of research became crucial when Farman et al. (1985) reported a massive stratospheric ozone decrease during polar spring in Antarctica. Today, the halogen chemistry of stratospheric ozone depletion seems to be well understood and has been summarized by several reviews (e.g. Solomon (1999); Rowland (2006)).

The driving forces behind stratospheric ozone depletion are catalytic halogen cycles, in which chlorine and bromine destroy ozone and are restored at the end of the cycles (Seinfeld and Pandis, 2006). While the chlorine mixing ratio in the stratosphere is about 3400 ppt, the bromine mixing ratio is only about 20 ppt. Nevertheless, bromine is approximately as important as chlorine. This is caused by two effects: a) Br is rapidly released by its precursor compounds and b) a major fraction of bromine exists in the highly reactive forms of Br and BrO. Further, the formation of different reservoir species like chlorine nitrate ($ClONO_2$) and HCl leads to temporary or permanent removal of chlorine from this catalytic cycles. A summed up, the bromine ozone-depletion potential is about 50 times higher than the chlorine (Seinfeld and Pandis, 2006).

1.5 Atmospheric halogen chemistry

While the only natural source for organo-chlorine species in the stratosphere is CH_3Cl (Rowland, 2006), which supplies about 500 ppt of total chlorine (Seinfeld and Pandis, 2006), anthropogenic emissions of chlorofluorocarbon compounds like CFC-113, HCFCs, CFC-11, CFC-12, and carbon tetrachloride complement the amount to the observed 3400 ppt of chlorine in the stratosphere. The contribution of naturally released species to the total content in the atmosphere is larger for bromine compared with chlorine. Bromoform and methyl bromide amount to about 25 % of the 20 ppt bromine in the stratosphere. Antropogenic bromine is emitted in the form of halons, like Halon-1301 and Halon-1211.

Halogens also seem to play an important role in tropospheric ozone destruction (Foster et al., 2001). Different atmospheric halogen species were monitored within the tropospheric boundary layer using differential optical absorption spectroscopy (DOAS) (Platt, 2000). Wagner et al. (2007) measured enhanced tropospheric BrO above the antarctic sea using MAX-DOAS (multi-axis DOAS). Enami et al. (2007) discussed the contribution of bromine, generated by iodine catalysis, to the global inorganic bromine in the troposphere. Apart from anthropogenic sources of halogen-organics, several natural sources for halogens and halogen-organics in the troposphere and the boundary layer were discussed. Important sources for halogens in the troposphere are halogen release from sea-salt aerosol (Finlayson-Pitts, 2003) and heterogeneous reactions on those aerosol surfaces (Rossi, 2003). Reaction schemes of these processes are described in detail, e.g. by Finlayson-Pitts (2010).

1.5.1 Halogen release from sea-salt aerosol and salt pans

With an estimated flux of about 10100 Tg y^{-1}, sea-salt aerosol constitutes the largest of the major aerosol classes (Gong et al., 2002). In general, sea salt consists of 55.04 wt.% chlorine, 0.19 wt.% bromine and 7.68 wt.% SO_4^{2-} as well as cations of the first and second group of the periodic table (Seinfeld and Pandis, 2006). On average, sea water contains about 3.5 % of sea salt. This value varies depending on the location, especially for isolated sites like the Dead Sea.

Sea-salt aerosol particles can be photo-activated to release gaseous halogen species. These effects were observable in early smog-chamber experiments of halogen release from $NaCl$ (e.g. Zetzsch et al. (1988); Behnke and Zetzsch (1990); Siekmann (2008)). Therefore, da Rosa (2003) measured equilibrium concentrations of various halogen species at different temperatures and pH values in water and salt solutions. Frinak and Abbatt (2006) report Br_2 production from the heterogeneous reaction of gas-phase OH with aqueous salt solutions. The tropospheric reaction mechanisms of halogen activation differ significantly from those in the stratosphere (Platt and Hönninger, 2003). Von Glasow and Crutzen (2004) summarized the complex reaction mechanisms and detailed the halogen release from different sources, supported by model

1 Introduction

calculations. The modeling of tropospheric halogen chemistry allowed a detailed insight into complex reaction mechanisms (Tas et al., 2006). The mechanisms of halogen release seem to play a major role in polar boundary-layer ozone depletion events (Simpson et al., 2007).

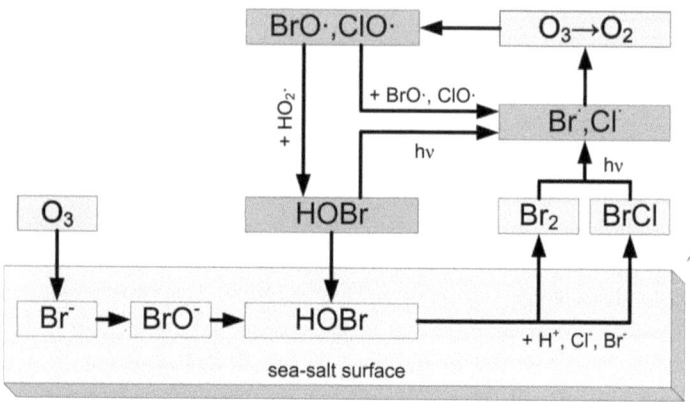

Figure 1.7: Simplified scheme of the halogen-release mechanism from sea-salt surfaces (according to Finlayson-Pitts (2010))

A simplified scheme of the complex heterogeneous halogen-release mechanisms is given in figure 1.7. The ozone-initiated oxidation of halogen species in a quasi-liquid (only one or a few molecular layers act as liquid layers) sea-salt-containing layer causes the release of gaseous Br_2 or, after depletion of bromide, $BrCl$. These species are photolyzed, and the subsequent formation of the oxidized species BrO, ClO and their HOX successors is accompanied by ozone depletion. $HOBr$ can be reabsorbed by the sea-salt surface and thus initiate another halogen-release cycle. Each molecule of $HOBr$ is able to release two Br atoms, which react to form two $HOBr$ molecules. Therefore, this cycle is also called "bromine explosion".

Recent aerosol smog-chamber studies reveal halogen release from $NaCl/NaBr$ salt pans with a wet surface (Balzer et al., 2010; Bleicher et al., 2010). Using DOAS (Perner and Platt, 1979), they report the formation of 10 ppb BrO and severe ozone depletion above the simulated salt pan in a Teflon smog-chamber after switching on the solar simulator (see figure 1.8). The release mechanism of halogens from the salt pan seems to be comparable to the mechanism of halogen release from sea-salt aerosol (Figure 1.7). Thus, tropospheric halogen species might also be found above continental salt lakes and brines. However, the composition of their salt layers might differ significantly from sea salt.

1.5 Atmospheric halogen chemistry

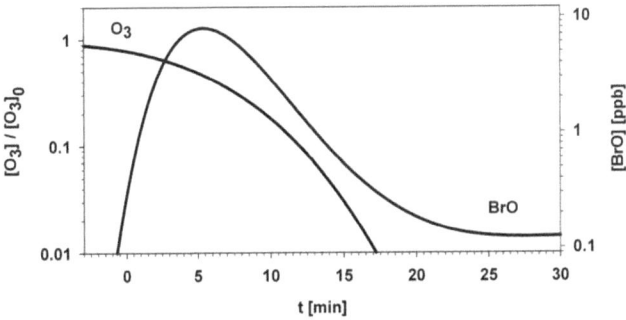

Figure 1.8: Halogen release from a simulated salt pan in a Teflon smog chamber (with kind permission of N. Balzer and J. Buxmann)

1.5.2 Halogen interaction with atmospheric gases, aerosols, and organics

The stratospheric halogen chemistry is limited to ozone-depleting catalytic cycles and the formation of reservoir species. Reactive halogen species are removed from the stratosphere by the formation of stable species, like HCl, and subsequent migration to the troposphere, which can lead to a total removal of those halogen species from the atmosphere (Seinfeld and Pandis, 2006). In the troposphere, halogens have more options, like reactions with other trace gases, with aerosol particles and adsorption on droplets. However, while sources of reactive halogen species in the troposphere are well characterized, their sinks are hardly known.

One possible sink is the reaction with organic matter, which is well known from water disinfection (e.g. Gallard and von Gunten (2002); Uyguner et al. (2004)). Trihalomethane formation from the reaction of molecular chlorine with humic substances was measured by Uyguner et al. (2004) in water-treatment processes. The formation of chloroacetic acids from soils, humic acids, and phenolic moieties was studied by Fahimi et al. (2003), suggesting a Fenton reaction mechanism. The importance of hydrophobic organics with phenolic acidity in producing high amounts of brominated organics was reported by Huang et al. (2004). More than 40 reaction products, such as chlorinated ketones, carboxylic acids, ketoacids, lactones, and furanones, were identified from the chlorination of orcinol (3,5-dihydroxy-toluene) (Tretyakova et al., 1994).

Rudich (2003) expects a concurrence of high concentrations of bromine and chlorine, released from marine aerosols, with organic aerosols in marine environments. Moise and Rudich (2001) used a low-pressure flow reactor to study the reactive uptake of Cl and Br by organic films.

1 Introduction

By using FTIR spectroscopy, they monitored an almost complete disappearance of aliphatic C-H bonds. Furthermore, they observed no halogenated reaction products on their surface films, but a formation of carboxyl groups, e.g. in carboxylic acids. Hydrogen abstraction and an enhancement of the hydrophilic character of the surfaces were reported (Rudich, 2003). Ishikawa et al. (1986) studied the reaction of chlorine and bromine with humic substances and found a post-bromination of chlorinated compounds. They concluded that bromine is more reactive towards humic substances than chlorine.

Smoydzin and von Glasow (2007) discussed the influence of organic surface films on the halogen release from sea salt aerosols. The amount of gas-phase bromine and chlorine decreased in the presence of an organic monolayer on sea-salt aerosols. Their results suggest that the chlorine release mechanism is more affected by the organics than by the bromine. A box-model approach was used by Toyota et al. (2004) to study the photochemical interaction between VOC and reactive halogens. While hydrogen abstraction is the dominant path for reactions of chlorine with alkenes, reactions of bromine with alkanes are dominated by addition, leading to brominated alkyl-peroxy radicals.

Small organo-halogens formed by haloform-type reactions were observed by Carpenter et al. (2005). Boyce and Hornig (1983) report the reaction pathways of trihalomethane formation by halogenation of dihydroxy-aromatic compounds acting as models for humic acids. The authors proposed an abiotic formation pathway via reaction of HOX with organic material on the quasi-liquid layer above the sea ice/snow pack. Kopetzky and Palm (2006) reported the formation of halogenated methanes and acetones by halogenation of humic acids in saline solutions. Bromoform, bromoacetones, and trihalomethanes are reported after ozonization of saline solutions containing significant amounts of humic acids (Sörgel, 2007). Caregnato et al. (2007) used flash-photolysis experiments to study the reactions of chlorine radicals with humic acids and reported rate constants up to $(3 \pm 2) \times 10^{10}$ mol^{-1} s^{-1}. They suggested that these high values could result from the interaction of $Cl_2^{\cdot-}$ with carboxylic acid groups of the humic acids.

While the atmospheric chemistry community takes close interest in heterogeneous processing reactions of atmospheric aerosols, (George and Abbatt, 2010), less effort has been focused on halogenated particulate matter or organic aerosols. Only a few field-measurement campaigns were reported. Rahn et al. (1979) measured halogen-containing particles as pollutants in New York, where Br is the most important halogen. Halogenated aerosols were reported by Mosher et al. (1993) and related to the photochemical aerosol production from biogenic organo-halogens. Cl-dominated particulate matter from anthropogenic sources, like pesticides, was measured by Xu et al. (2005) at a downtown site in Beijing. Holzinger et al. (2010) found halogenated compounds in organic aerosols at the Sonnblick observatory (Austria), but were not able to identify them.

1.6 Scope of this work

The aim of this study is to investigate heterogeneous interactions of halogens released from natural sources with different types of organic aerosols. A simple interaction scheme is given in figure 1.9.

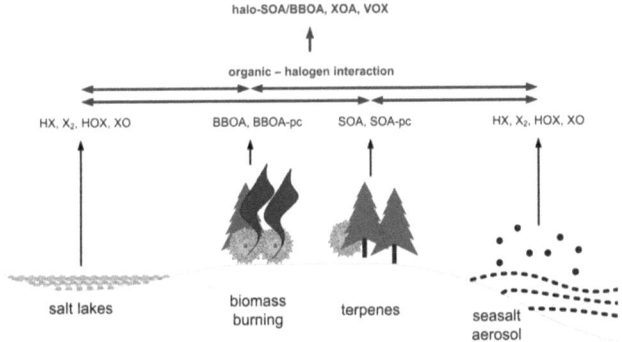

Figure 1.9: Possible formation of halogenated SOA or BBOA (halo-SOA/BBOA), secondary halogen-organic aerosol (XOA), and halogenated organic gaseous species (VOX) by interaction of reactive halogen species released from salt lakes or sea-salt aerosol with SOA, BBOA, or their respective organic precursors (SOA-pc, BBOA-pc).

Several locations exist, where those interactions can be expected. Two examples are given in figure 1.10: a salt lake in West Australia, surrounded by possibly terpene-emitting biomass and influenced by biomass burning, and a coastline in Croatia, where terpene emissions occur in the vicinity of sea-spray aerosol.

Molecular halogens and simulated natural sources of halogens (salt pan and sea-salt aerosol) were used to study those interactions on a laboratory scale. The formation and transformation of different organic aerosols and gaseous species reacting with halogen species were monitored.

While the analysis of single compounds of the organic aerosols is futile because of the large amount of different compounds, spectroscopic methods were used to monitor physicochemical changes in the particulate matter of the organic aerosol. Some methods had to be developed or adapted to obtain detailed information on the chemical state of the aerosol (see chapter 2). Since no adequate model substance was available to study reactions of atmospheric HULIS

1 Introduction

(a) Coastline of the island of Rab (Croatia): emissions of terpenes close to sea-spray aerosol, emitting reactive halogen species

(b) Salt lake close to Lake King (West Australia): terpene emissions and biomass burning around salt lakes

Figure 1.10: Examples of remote areas where organic-halogen interaction might take place.

in an aerosol smog chamber without the need of complex sample preparation techniques, two aromatic precursors (catechol and guaiacol) were chosen to form SOA, simulating the olefinic or the aromatic core of atmospheric HULIS.

The following steps were carried out to be able to verify single changes of aerosol parameters during the heterogeneous halogen-aerosol interaction:

1. Appropriate SOA precursors were chosen to cover a wide range of aliphatic and aromatic features of organic aerosols, especially of HULIS.

2. SOA formation from these precursors was studied using an aerosol smog chamber under varying environmental conditions (simulated sunlight and relative humidity).

3. The particulate matter of the resulting SOA was characterized comprehensively. In case of SOA from catechol and guaiacol also the formation process was studied.

4. The halogen-aerosol interaction of the characterized organic aerosols was studied using molecular chlorine or bromine, which were photolyzed by UV/VIS irradiation.

5. Finally, the different types of SOA were exposed to natural sources of gaseous halogens to test the halogen-aerosol interaction, to determine the physicochemical transformation of the particulate matter and the influences on the halogen release mechanisms.

2 Methods

A variety of methods is available to study surfaces of airborne particles, like aerosols, and their interaction with atmospheric trace gases (Nießner, 1991; Zellner et al., 2009). Surface functional groups play an important role for heterogeneous aerosol chemistry (Lary et al., 1999). For studying the formation and processing of those functional groups, FTIR spectroscopy is most suitable (Najera et al., 2009). Time-resolved FTIR spectroscopy monitoring the formation of organic aerosol particles was applied by Sax et al. (2005). The FTIR-spectroscopic investigation of secondary organic aerosol formation in an aerosol smog chamber provides a deeper understanding of these processes (Sax et al., 2005). SOA in aerosol smog chambers can be formed at conditions resembling those in the atmosphere with respect to ozone and precursor concentrations, relative humidity, or UV irradiation. Coury and Dillner (2008) used attenuated total reflection (ATR)-FTIR spectroscopy to quantify organic functional groups in ambient aerosols. Furthermore, light absorption of organic materials plays an important role for the understanding of radiative forcing (Shapiro et al., 2009).

To prepare the organic aerosols and study their interaction with halogen species, aerosol smog chambers and an aerosol flow reactor were used. The organic aerosols were formed as secondary aerosols in aerosol smog chambers (see section 2.1) or in an aerosol flow reactor (see section2.2). Formation studies and processing with molecular halogens were performed using a 700 L glass chamber (see section 2.1.1). The interaction of secondary organic aerosols with naturally released halogens were studied using a 3500 L Teflon chamber (see section 2.1.2).

The aerosol size distribution of the processed organic aerosols was determined using an electrostatic classifier coupled to a condensation nucleus counter (see section2.3). Imaging of the solid phase of SOA from catechol and guaiacol by electron microscopy was done after collecting the particles on filters (see section 2.4). The chemical structure of the particulate matter was studied using infrared spectroscopy (see section 2.5). Long-path absorption FTIR allowed to study the aerosol formation and processing in the 700 L glass chamber (see section 2.5.1). The particulate phase was analyzed using ATR-FTIR spectroscopy (see 2.5.2). Two electrostatic precipitators were developed to deposit the particulate phase onto the ATR crystals (see 2.5.3). Temperature-programmed pyrolysis mass spectroscopy (TPP-MS, see section 2.6) and ultra-high-resolution mass spectroscopy (see section 2.8) were used to detect functional groups and

2 Methods

to analyze halogenated organic species as well as O/C and H/C ratios and the carbon oxidation state. The optical properties in the UV/VIS range were measured using diffuse-reflectance UV/VIS spectroscopy (see section 2.7).

2.1 Aerosol smog chambers

2.1.1 700 L glass chamber

The general setup of this glass chamber has been described by Nolting et al. (1988). However, due to modifications after relocation of the chamber to Bayreuth and an increase of the volume of the chamber by adding another 60 cm section, the setup has significantly changed (Fig. 2.1).

The glass chamber now consists of four cylindrical elements made of Duran glass. The chamber has an overall height of 2.3 m and a diameter of 0.6 m. The top and the bottom of the aerosol smog-chamber are sealed by fluorinated ethylene propylene films (FEP 200A, Dupont). These membranes also serve as windows for transmitting UV/VIS radiation emitted by the solar simulator. The four glass elements are equipped with one to four flanges each. In total, the chamber has nine flanges to mount instruments, tubes for sampling or supplying gases, or precursor inlet systems. The volume of the glass chamber in its current setup, including the flanges, was calculated to be 680 L, and the surface-to-volume ratio is 8.7 $m^2 m^{-3}$.

At the bottom, the glass chamber is equipped with a medium-pressure metal vapor lamp (Osram Metallogen HMI, 4000 W) as a solar simulator. A water-cooled glass disk cuts off the UV-C range of the lamp spectrum (Fig. 2.2). The about one centimeter thick water layer reduces the amount of infrared radiation from the HMI lamp entering the smog chamber. The lamp spectrum of the solar simulator was measured using a Bentham M300HRA monochromator, calibrated using a calibration lamp (Model Oriel 63361), with a 1P28 photomultiplier. A glass reflector was used to direct the light into the monochromator. According to figure 2.2, the spectral actinic flux, with the UV radiation cut off at about 290 nm, is close to the calculated as well as to the reported solar spectral actinic flux (Seinfeld and Pandis, 2006). The calculation of the solar spectrum shown in 2.2 was done using the software package STARsci for the coordinates 40° N 11° E, summer, maritime albedo, 348 DU of ozone, and no aerosol. While the black-body temperature of the calculated solar spectrum is 5409 K, the temperature of the solar simulator is 3940 K without and 3870 K with the UV-C filter. Hence, the maximum of the spectrum is shifted to higher wavelengths compared to the solar spectrum.

The glass chamber is permanently flushed with particle-free air supplied by a zero air (dry and contaminant-free air, which is suitable for use as a zero reference calibration gas) generator (CMC ZA 50K for SOA formation studies and CMC ZA 100K for halogen interaction studies)

2.1 Aerosol smog chambers

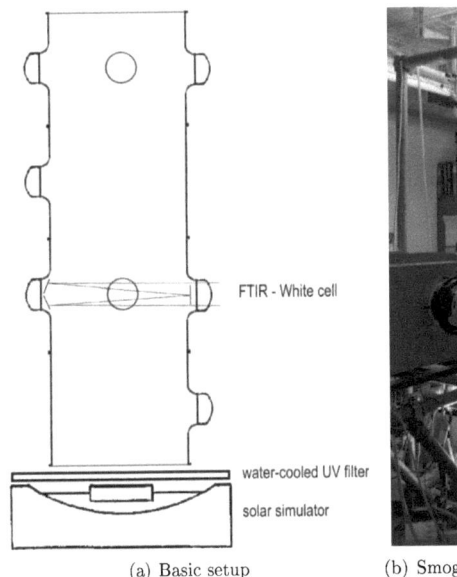

(a) Basic setup

(b) Smog chamber with solar simulator in operation

Figure 2.1: 700 L aerosol smog chamber

and kept at an overpressure of 5–10 Pa. The dew point of water is decreased to -70 °C (5.18 ppm), and the concentration of CO_2 is reduced using a Balston 60-75 dehumidifier. The overpressure was also maintained during the experiments. With the solar simulator in operation, residence times of 105±15 min were measured for particles with diameters of 50±5 nm.

The ozone concentration in the chamber was monitored using a chemiluminescence ozone analyzer (UPK 8002). Pure oxygen (Riessner-Gase, > 99.995 %) was used to prepare ozone using a silent-discharge ozonizer (Sorbios, GSG 0012).

The relative humidity in the smog chamber was adjusted at the beginning of the experiments by vaporizing a calculated amount of double-distilled water. Due to the fact that all experiments were performed at room temperature (approximately 25 °C), the saturation vapor pressure of water was calculated using the Magnus equation (2.1) as reported by Sonntag (1990).

$$E_w = 6.112 \times exp(\frac{17.62t}{243.12 + t}) \tag{2.1}$$

2 Methods

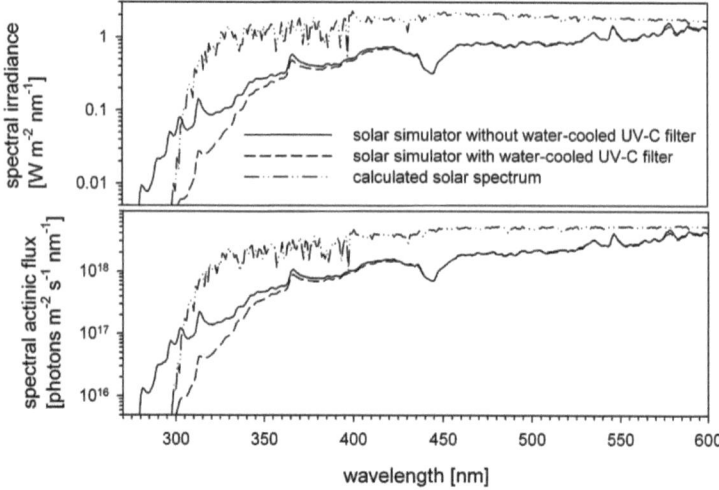

Figure 2.2: Spectrum of the solar simulator Osram Metallogen HMI 4000 W with and without the water-cooled UV-C filter, compared to a calculated solar spectrum (40° N 11° E, maritime albedo, summer, no aerosol, 348 DU of ozone)

By equation 2.1 the saturation vapor pressure of water E_w in hPa is calculated at the temperature t in °C. The ideal gas law gives access to the saturation mass of water $\rho_{w,max}$ in gm^{-3} (2.2), where R_w is the specific gas constant of water (461.52 Jkg^{-1}K^{-1}).

$$\rho_{w,max} = \frac{100 E_w}{R_w(t + 273.15)} \qquad (2.2)$$

Finally, the necessary amount of water to achieve a given relative humidity $\rho_{w,SC}$ in g m^{-3} is derived from the definition of the relative humidity (RH) (Seinfeld and Pandis, 2006) and calculated for the volume of the smog chamber V_{SC} (2.3).

$$\rho_{w,SC} = \frac{RH}{100} \rho_{w,max} V_{SC} \qquad (2.3)$$

The relative humidity in the aerosol smog chamber was monitored using a Steinecker hydrometer (Type 49076D). The use of long-path infrared absorption spectroscopy (for details see subsection 2.5.1) is limiting the applicable relative humidity to 25 % and therefore the maximum amount of water to 4.25 g. The injected amount was 5 g, allowing for dilution of the water vapor by flushing the aerosol smog chamber.

2.1.2 3500 L Teflon chamber

The interaction of SOA with halogens released from a simulated salt pan was studied using a 3.5 m^3 Teflon smog chamber. FEP 200a Teflon film (Dupont) was chosen as wall material. The initial setup was assembled and described in detail by Siekmann (2008). Seven HMI lamps with 1200 W each are used as a solar simulator. To cut off the UV-C band of the spectrum, a water-cooled glass disk (Tempax glass, Schott AG) was installed.

An FEP sheet carrying a thin layer of salt grains was installed in the lower third of the Teflon chamber to simulate a salt pan as a source of atmospheric halogen species for the processing of the organic aerosol.

2.2 Aerosol flow reactor

Time-resolved long-path absorption FTIR has one major limitation, which is the time span to measure a single spectrum. Typically, 64 to 512 interferograms are combined to calculate a single spectrum with a suitable signal-to-noise ratio. To measure those interferograms at resolutions between 1 and 4 cm^{-1}, several minutes with a Bruker IFS 113v FTIR spectrometer (for details see section 2.5) are required. As a result, every spectrum represents an average of this time span.

The essential step during SOA formation is the reaction of the SOA precursor with atmospheric trace gases. This interaction leads to the transformation of the precursor molecule and to nucleation, chemical bonding, and particle formation. The timescale for this reaction is too short to allow for it being monitored by FTIR spectroscopy in aerosol smog chambers equipped with infrared White cells. If the precursor concentrations need to be high because of the detection limit, the transformation of the aerosol precursor is even faster.

Aerosol flow reactors (AFR) are very suitable for studying heterogeneous reactions of aerosols (Kodas et al., 1986; Remorov et al., 2005). Application of optical methods for UV/VIS (Vesna et al., 2009) and infrared spectroscopy (Najera et al., 2008) gives information on trace gases, reaction products, and the particulate fraction of aerosols. The usage of preformed aerosols (by atomizing, ultrasonic nebulizing, or homogeneous nucleation) to study heterogeneous reactions with atmospheric traces gases in aerosol flow reactors was already reported (Najera et al., 2009; Last et al., 2009).

However, aerosol flow reactors also permit to study aerosol formation processes during the first few seconds after mixing of the precursor with the reacting atmospheric trace gases, like ozone. Particles arriving at the absorption cell were all formed at the same time. The content

2 Methods

of the absorption cell thus represents a temporally well-defined state of aerosol formation and processing, which depends on flow profile, speed, and length of the aerosol flow reactor.

Postulating a laminar flow, the temporal resolution of the absorption cell is only limited by the residence time of the aerosol inside the cell. By changing the length of the reaction zone, the age of the aerosol entering the measuring cell can be varied.

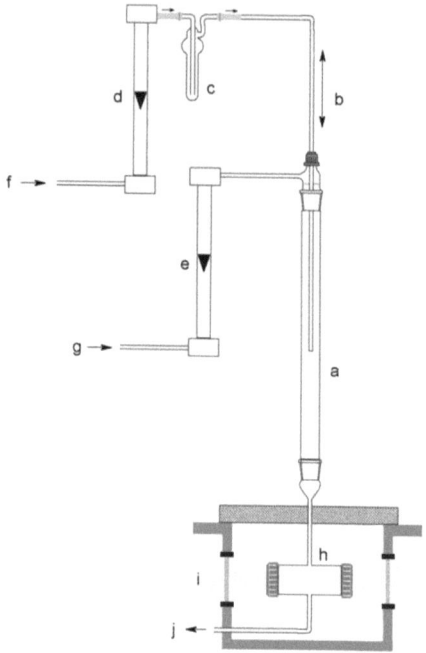

Figure 2.3: Setup of the aerosol flow reactor coupled to the FTIR spectrometer: a–flow reactor; b–movable inlet; c–impinger to vaporize the aerosol precursor; d & e–flow meter to control the gas flows; f–carrier gas inlet; g–reactive gas inlet; h–infrared gas cell; i–FTIR sample compartment; j–outlet.

To study the first seconds of aerosol formation, an AFR was constructed (Figure 2.3). The AFR allowed to record infrared spectra at a defined time (between 1-10 seconds) after the mixing of the precursor with the oxidant.

The front port of the Bruker IFS 113v FTIR spectrometer is equipped with an atmospheric pressure sample chamber, which is separated from the vacuum of the spectrometer by two KBr windows. The gas absorption cell is equipped with KBr windows and has a volume of 22.6 cm^3. It was placed in the center of the infrared beam and connected to the AFR, which has a length of 50 cm (Figure 2.3). Inside the AFR (inner diameter 2.2 cm), an adjustable glass inlet (inner diameter 2.6 mm) is centered. The AFR is arranged vertically to reduce particle deposition on the walls through gravitational settling and upwelling/thermal convection caused by the heat of the reaction. The upper end of the AFR is connected to a flow meter (Rota) with a needle valve to control the ozone flow. The adjustable inlet is connected to a gas saturator to adjust the precursor concentration and to another flow meter (Rota) and needle valve for the carrier gas.

The particle formation process is started by adding the precursor gas flow to the ozone-enriched oxygen flow under isokinetic conditions. The flows of oxygen (1 Lmin^{-1}) and precursor carrier gas (20 to 100 cm^{3}min^{-1}) lead to an overall residence time of about 11 s inside the AFR and about 1.3 s inside the gas cell. The ratio of the gas flows is varied between 1:20 and 1:50. Aerosol formation can be monitored between 1 and 10 seconds after mixing the two gas streams by changing the position of the gas inlet in the AFR. The whole setup and the flows were adjusted to the low Reynolds-number range to achieve laminar flow conditions.

The setup and first results of the aerosol flow-reactor experiments are published by Ofner et al. (2010a).

2.2.1 Circular multi-reflection cell

Based on the experience with the aerosol flow reactor, a multi-reflection cell with an optimized geometry for the use with a flow setup was developed. The main existing concepts of multi-reflection cells, developed by White (1942) and Herriott et al. (1964), are characterized by reflecting the beams along a special axis of the optical setup. A review of those setups has been given by Robert (2007). The optical paths of those setups are not focused on a special point. Hence, a complete and homogeneous filling of the space within the measuring cell is needed to exploit the full length of the optical path.

To couple an aerosol flow reactor with a long-path optical device for infrared spectroscopy with improved detection limits, very small White-cell optics have been used (Najera et al., 2008). But these cells come up with some disadvantages: Apart from the difficulty to adjust those devices inside the experimental setup or the vacuum system, only a small fraction of the overall beam is passing the stream of precursor or aerosol centered inside the cell. Hence, a circular multi-reflection cell, which focuses the infrared beam in or close to the center of the cell, appears to be more appropriate. Such a multi-reflection cell was constructed using separate mirrors, and it allows for an increase in path length by a factor of 8 (Thoma et al., 1994). However, the path length can not be varied, and there is still the need of adjusting every single mirror inside the experimental setup.

2 Methods

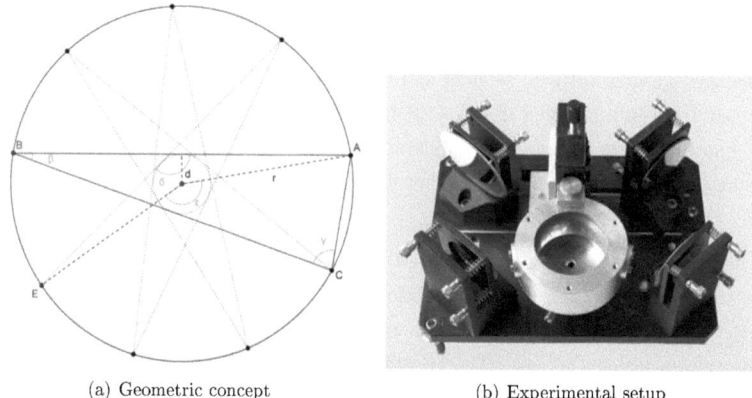

(a) Geometric concept   (b) Experimental setup

Figure 2.4: Geometric concept and setup (photo by Christian Wißler, University of Bayreuth) of the circular multi-reflection cell

The basic geometric concept of the circular multi-reflection (CMR) cell is shown in figure 2.4(a). The cell is equipped with two apertures for the beam to enter and to leave at positions A and E, defining the angle ϵ. The beam entering at A is reflected at position B. The angle of reflection β is dependent on the offset d of the two apertures from the center of the cell. After multiple reflections inside the cell, the beam leaves the cell at position E.

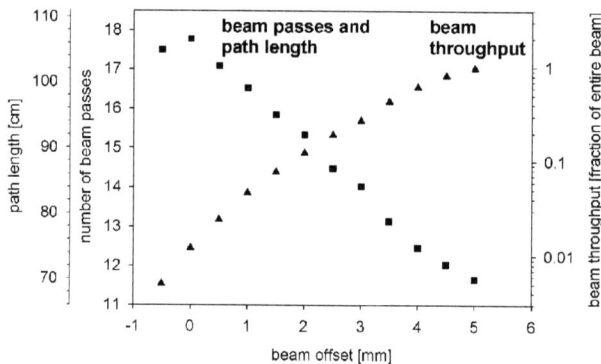

Figure 2.5: Path lengths achieved with the circular multi-reflection cell

A detailed description of the circular multi-reflection cell including the mathematical calculations of achievable beam lengths, the geometric concept, and the basic design is given by Ofner et al. (2010b). Based on those concepts, the circular multi-reflection cell was manufactured at the University of Bayreuth. The CMR cell consists of an aluminum cylinder with an outer diameter of 80 mm and a height of 30 mm, with a polished spherical reflecting inner surface, focusing the light beam on the opposite walls each. The radius r of the concave mirror is 30 mm. The cell is mounted on a standard optical bench (Bruker Optics) with four mirrors focusing the infrared beam from the interferometer compartment into the cell and from there into the detector compartment of the spectrometer (Figure 2.4(b)).

The performance of the cell was characterized using CO_2 (see Ofner et al. (2010b) for details). The number of theoretical reflections inside the circular multi-reflection cell reached 17.5, and hence path-lengths of up to 105 cm could be achieved (Figure 2.5). The focusing of the infrared beam close to the center of the cell is ideal for using this cell coupled to the aerosol flow reactor described above.

2.3 Particle sizing

All aerosol size distributions were measured using an electrostatic classifier (TSI 3071) with a particle neutralizer (^{85}Kr; TSI 3077A) coupled to a condensation nucleus counter (TSI 3020).

The calculation of the particle diameters from particle mobilities and the normalization of the particle concentration per channel were done as indicated by Knutson and Whitby (1975). The aerosol size distribution was obtained by correcting the electrical charge of the particles with the multi-polar charge distribution described by Wiedensohler (1988). All presented medians of the particle diameters $\bar{d}_p$ were derived by fitting all single aerosol size distributions with the log-normal distribution (Reist, 1993). Based on the log-normal distribution, the geometric standard deviation σ_g was calculated to describe the shape of the distribution.

Aerosol size distributions were measured with a temporal resolution of 8 minutes, resulting from by the design and setup of the system. Each aerosol size distribution was scanned in both directions with respect to particle diameters.

2.4 Electron microscopy of SOA

Electron microscopy of unreacted particles of SOA from catechol was performed at the Vienna University of Technology using a FEI Quanta 200 field-emission-gun scanning electron microscope (FEG-SEM). The particulate matter of the aerosol was collected using $Isopore^{TM}$ membrane filters (Millipore), made of polycarbonate, with a pore size of about 50 nm. Another sample preparation technique was the precipitation of the particulate matter on Si wafers using a home-made electrostatic precipitator (see section 2.5.3).

2 Methods

The particles were protected against electrical charging and subsequent destruction by the electron beam by sputtering 3–4 nm of Au/Pd onto their surface. The FEG-SEM was operated at 6×10^{-6} mbar using a cathode voltage of 5 kV to avoid high penetration depth and only image the surface and morphology of the cluster and chain structures of the samples.

2.5 Fourier-transformation infrared spectroscopy (FTIR)

Two Fourier-transformation infrared (FTIR) spectrometer were used to study the formation and transformation of structural elements and oxygen- or halogen-containing functional groups of the model SOA.

A high-resolution Bruker IFS 113v spectrometer is coupled to the 700 L aerosol smog chamber (for details see section 2.5.1). While the back port of the IFS 113v is used for transmitting the beam coming from the smog chamber through the interferometer to the detector compartment, the front port of the spectrometer is equipped with an atmospheric-pressure sample port for Bruker standard optical benches. This port was used for the aerosol flow-reactor setup (see section 2.2). The Bruker IFS 113v is equipped with a DTGS (deuterated triglycine sulfate) detector and a mid-band MCT (mercury cadmium telluride semiconductor) detector. A globar is installed in the source compartment of the spectrometer and used as light source for the mid-infrared spectral region. For the same spectral region, a KBr/Ge beam splitter is installed in the interferometer compartment. The interferogram is generated using a Genzel-type interferometer (for details see Griffiths and Homes (2001)). Acquisition of all spectra was done in the range of 8000 to 0 cm^{-1}, using a high-pass filter of 1.1 kHz cutting off at 585 cm^{-1} and a low-pass filter of 12.5 kHz cutting off at 6642 cm^{-1}. The acquisition mode was single sided using no spectral correlation. For the Fourier transformation, a phase resolution of 16 was chosen using the Mertz method for phase correction. The Blackman-Harris 3-term method was used for apodization with a zero-filling factor of 2. Generally, 512 single interferograms were recorded at a resolution of 2 cm^{-1} and a mirror velocity of 29.73 kHz. All spectra were evaluated between 4000 and 600 cm^{-1}.

A low-resolution Bruker IFS 48 spectrometer was used for analyzing the particulate matter of the organic aerosol using ATR-FTIR spectroscopy (see section 2.5.2). The spectrometer is equipped with a standard Michelson interferometer, a DTGS detector, a globar mid-infrared source, and a KBr/Ge beam splitter. For each spectrum, 512 single interferograms were recorded between 4000 and 400 cm^{-1} at a resolution of 4 cm^{-1} and a mirror velocity of 10 kHz. Acquisition was also performed between 4000 and 400 cm^{-1} in single-sided forward-backward mode. The full interferogram length option was chosen as correlation mode. The Fourier transformation was performed at a phase resolution of 64, and Mertz-signed as phase

2.5 Fourier-transformation infrared spectroscopy (FTIR)

correction method. All other parameters were set to the same values as for the Bruker IFS 113v.

Both spectrometers were operated with particle-free zero air, as described in section 2.1.1. Post-processing and atmospheric compensation of the infrared spectra was performed using the Bruker Opus software package (version 5.0). For details on the chosen settings for the two FTIR spectrometer described above, see Davis et al. (2001); Griffiths and de Haseth (2007).

2.5.1 Long-path infrared absorption spectroscopy in the 700 L glass chamber

The 700 L aerosol smog chamber was equipped with a 40 m White cell (White, 1942) designed by Bruker Optics and coupled to the FTIR spectrometer (Bruker IFS 113v). The spectrometer and the connection to the smog chamber were evacuated to 60 mbar to reduce disturbances of the spectra by atmospheric compounds.

The White cell was adjusted by coupling a He-Ne laser with the back port of the spectrometer, through the interferometer, the source compartment, the White cell, and finally to the separated source compartment equipped with a globar. The source compartment was connected to the spectrometer by a glass tube. For adjusting the White cell, the beam splitter was replaced by a Mylar-6ae film for allowing the He-Ne laser beam to pass.

The path length L_W of the White cell is given by equation 2.4.

$$L_W = 2L_b(1+2R) \qquad (2.4)$$

The cell with a base length L_b of 80 cm is optimized for 12 double reflections (R) with a corresponding optical path length of 40 m. The system could also be optimized for up to 18 reflections, resulting in optical path lengths of up to 60 m.

The three mirrors of the White cell had been coated with gold by the optics workshop of the University of Bayreuth using a physical vapor deposition (PVD) process.

2.5.2 Attenuated total reflectance (ATR) spectroscopy

FTIR spectroscopy can provide structural and chemical information on organic macromolecules such as components of atmospheric aerosols. However, these samples exhibit high optical density and therefore, in situ transmission measurements are often hampered by an insufficient signal-to-noise ratio, while transmission spectroscopy using KBr pellets is much more time

2 Methods

consuming (Havers et al., 1998). An alternative strategy is provided by reflection techniques, since they have the inherent advantage of being independent of sample thickness, and unlike transmission spectroscopy, they avoid the influence of light scattering on the spectral data, which is a source of interference in absorption spectroscopy of opaque samples (Muckenhuber and Grothe, 2007).

ATR spectroscopy is a powerful technique to analyze secondary organic aerosol particle fractions using FTIR. Such spectra are reported in the literature for various secondary organic aerosols (Dekermenjian et al., 1999). To perform ATR spectroscopy of organic aerosols, various sample preparation techniques have been described in the literature. Most common is the sampling of atmospheric compounds on filters. These samples are then transferred to the ATR crystal by impression (Ghauch et al., 2006). Quantitative measurements are only possible with limited reproducibility because of incomplete transfer of the sample from the filter onto the crystal. Furthermore, the potential contribution of the filter material to the infrared spectrum can lead to incorrect interpretation. A way to avoid this would be the use of an impactor to collect atmospheric aerosols directly on the ATR crystal (Allen et al., 1994). This technique allows size-dependent measurements, but will not allow to analyze the total particle fraction (Johnson et al., 1983). Direct deposition of model aerosol samples by gravitational deposition onto ATR crystals yielded promising results (Zhang et al., 2005). Hence, an electrostatic precipitation technique was developed to deposit the particulate matter of the organic aerosols onto the ATR crystals, which is described in detail in section 2.5.3.

In general, the ATR spectra were recorded using a Bruker IFS 48 FTIR instrument with a Specac 25 reflection ATR optics. Three different types of ATR crystals with trapezoidal shape (52x20x2 mm) were available: Germanium (spectral range: 5000–400 cm^{-1}), Zinc Selenide ZnSe (spectral range: 20000–500 cm^{-1}), and Thallium bromide and iodide KRS-5 (spectral range: 16000–200 cm^{-1}) A detailed description of the spectroscopic setup is given above (see section 2.5).

2.5.3 Electrostatic precipitators

Morrow and Mercer (1964) described a single-stage point-to-plate electrostatic precipitator (ESP) for electron microscopy using field charging to gather and deposit aerosol samples. Some other designs of electrostatic precipitators were reported in literature. Mainelis et al. (2002) describe a precipitator for bio-aerosol collection on Agar plates, consisting of three components with separated charging and precipitation. Fierz et al. (2007) developed a portable ESP for TEM, also with separated charging and deposition. Such a direct deposition technique for

2.5 Fourier-transformation infrared spectroscopy (FTIR)

atmospheric aerosols on even surfaces appears to be a promising alternative to other methods of in-situ spectroscopy.

Based on those concepts, an electrostatic precipitator was developed to perform ATR-FTIR spectroscopy of organic aerosols from the aerosol smog chamber (see section 2.1). The concept is based on a point-to-plane ESP of the Rochester design (Abdel-Salam et al., 2007). It is a single-stage electrostatic precipitator in which one electric field will do both: charge the aerosol particles and deposit them on the crystal. This design was chosen for a convenient matching of the precipitation area to the crystal geometry, and because of the simple setup of the electric system (Figure 2.6(a)).

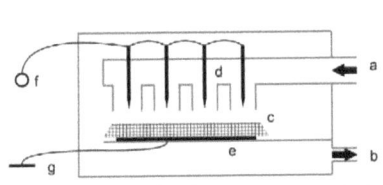

(a) Concept: a–inlet, b–outlet, c–crystal, d–needles, e–copper plate, f–high voltage supply, g–electric ground

(b) Experimental setup: ESP version 1 with a KRS-5 ATR crystal on the deposition plate

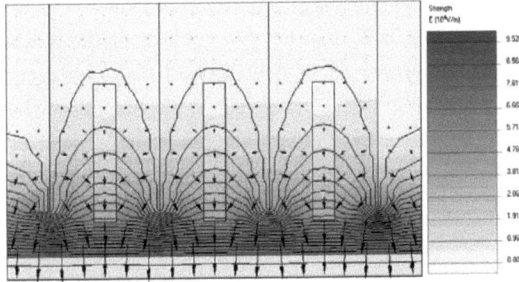

(c) Calculation: The maximum of the electric field strength is directed towards the crystal surface where deposition of the charged particles takes place.

Figure 2.6: Concept and picture of the experimental setup and calculation of the electric field of the electrostatic precipitator for coating ATR crystals–version 1

The ESP is situated in a cylindrical glass vessel (70 mm diameter, 160 cm length), wherein the ATR crystal is placed on a sample carrier made of epoxy resin. A copper layer on the sample carrier is connected to the electric ground. Four metal needles (curvature radius of the tip about 30 µm) are mounted inside the PVC inlet tubes pointing towards the ATR crystal

33

2 Methods

Table 2.1: Applicability of ATR crystal materials for the two electrostatic precipitators constructed

ATR crystal materials[a]	IR spec. range [cm^{-1}]	ESP v1 [kV][b]	ESP v2 [kV][b]
Germanium	5000 - 600	2.8	not tested
ZnSe	20000 - 500	4.0	10.5
KRS-5	16000 - 200	not applicable [1]	10.5

[a] delivered by Korth Kristall GmbH
[b] High voltage resulting in 99 % deposition efficiency
[1] KRS-5 crystal surface was eroded by the corona discharge.

surface. The needles are surrounded by the PVC tubes guiding the split aerosol flow towards the crystal surface, thus covering most of the surface of the ATR crystal with the full amount of the present aerosol particulate matter. The material of the needles is plain steel with no special characteristics. In normal operation mode the needles are about 1 mm above the crystal surface. The setup of the ESP is shown in figure 2.6(b).

The four needles are connected to a high-voltage power supply generating an electric field, and thus electrons/ions by corona discharge. The optimum voltage for depositing particulate matter onto ZnSe crystals was found to be 4 kV. The electrostatic conditions inside the ESP were calculated using the software package Student's Quickfield 5.8 (Tera Analyses Ltd.; see figure 2.6(c)). The calculation indicates that the gradient of the maximum field strength is directed towards the crystal surface where deposition of the charged particles takes place. A detailed description of this first version of the ESP, the deposition behavior depending on the permittivity of the crystals used, and the flow conditions applied is given by Ofner et al. (2009). Furthermore, ATR spectra from electrostatically deposited particles were compared to ATR spectra of filter samples, indicating no chemical transformation caused by the charging process of the electrostatic precipitation (Ofner et al., 2009).

The direct point-to-plate concept, which could be applied for ZnSe crystals without trouble, led to erosion of the softer material of the KRS-5 crystals by electrons emitted from the corona discharge. Hence, a second version of the electrostatic precipitator was developed. The precipitator is based on the concepts of Fierz et al. (2007) and Mainelis et al. (2002). This ESP has separate charging and deposition zones. The ESP is operated at a high voltage of about 10.5 kV. After entering the ESP, the aerosol particles are charged by the corona and accelerated towards a bevelled deflector at 3.5 kV. The high voltage drop of 7 kV between the deflector and the grounded copper layer turns the aerosol flight path towards the ATR crystal. The basic concept is shown in figure 2.7. A photography of the experimental setup is given in figure 2.7(b).

2.6 Temperature-programmed pyrolysis mass spectroscopy

The second version of the ESP has been described in detail by Ofner et al. (2011). The electric field strength was again calculated using the Student's Quickfield software package (see figure 2.7(c)). Due to the use of the bevelled deflector, the electric field strength is decreasing towards the crystal surface. Hence, erosion of the KRS-5 crystal is prevented.

The two ESP were operated in the positive mode of the high-voltage power supply, resulting in reduced ozone formation and a softer ionization. The applicability of the available ATR crystal materials for the two ESP is shown in table 2.1.

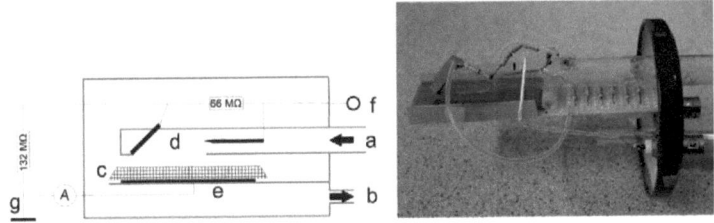

(a) Concept: a–inlet, b–outlet, c–crystal, d– needle and bevelled deflector, e–copper plate, f–highvoltage supply with cascade, g–electric ground

(b) Experimental setup: ESP version 2 with a KRS-5 ATR crystal on the deposition plate

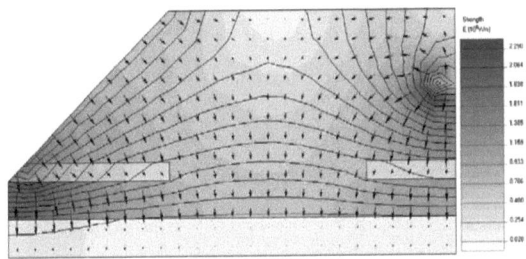

(c) Calculation: The use of the bevelled deflector allows to turn the flight path of the charged particles towards the crystal surface, while the overall field strength at the crystal surface is decreased.

Figure 2.7: Concept and picture of the experimental setup, and calculation of the electric field inside the electrostatic precipitator for coating ATR crystals–version 2

2.6 Temperature-programmed pyrolysis mass spectroscopy

Temperature-programmed pyrolysis mass spectroscopy (TPP-MS) is used to study functional groups on carbonaceous surfaces like graphitized carbon and soot (Dandekar et al., 1998). Fur-

2 Methods

thermore, processing of those surfaces with reactive atmospheric trace gases was studied by investigating the transformation of those surface functional groups (Muckenhuber and Grothe, 2004; Ofner and Grothe, 2007). The main structural elements of soot are composed of large aromatic structures establishing a large π-electron system. The functional groups, anchor points for heterogeneous reactions are located at the edges of this system. To study the amount and diversity of functional groups on the surface of soot without measuring the overall aromatic system, their different thermal stability was used by detecting the temperature-dependent desorption of molecules from the original functional groups.

The same idea has been transferred to the analyses of the functional groups of SOA. The technique seems to be suitable for organic aerosols with a very low vapor pressure (Jimenez et al., 2009), e.g. aged SOA or HULIS. Aerosol particles were sampled onto Whatman QMA quartz fibre filters (25 mm diameter). Particulate matter was collected out of 400 L of air from the glass chamber (see section 2.1.1) and 1000 L of air from the Teflon chamber (see section 2.1.2).

In TPP-MS experiments, the coated quartz fibre filters were placed in a quartz glass flask, which was then evacuated to 10^{-5} bar. Subsequently, the sample was heated in high vacuum from room temperature to 900 °C with a heating rate of 10 $Kmin^{-1}$.

Samples of the aromatic aerosol from catechol and guaiacol were measured using the TPD-MS system of H. Grothe (Institute of Materials Chemistry, Vienna University of Technology). Through a leak valve a small portion of the pyrolysis gases was introduced into the mass spectrometer (Balzers Prisma 200 QMS). The setup of the Vienna system has been described by Ofner (2006). Several fragment masses were recorded as a function of the pyrolysis temperature. To identify decomposing functional groups, the mass signals of OH (m/z=17), CO (m/z=28) and CO_2 (m/z=44) were exploited. Peaks were assigned to the decomposing functional groups (Muckenhuber and Grothe, 2006).

TPP-MS spectra of the halogenated organic aerosols and SOA from α-pinene were measured using a new setup at the Atmospheric Chemistry Research Laboratory in Bayreuth. The vacuum system of this TPP-MS system is shown in figure 2.8.

Similar to the Vienna system, this TPP-MS system consists of a tube furnace to pyrolyze the aerosol sample and a quadrupole mass spectrometer to analyze desorption and pyrolysis products. At this setup, all desorbing molecules were transfered directly to the MS section without using a leak valve. The QMS section was evacuated to less than 10^{-8} mbar using a turbomolecular pump (P 1.2) coupled to a membrane pump (P 1.1), the pressure was monitored using a full-range vacuum gauge (P(FR)). The ultra-high vacuum of the QMS section is separated from the high-vacuum system by a UHV corner valve (V1.1). A PrismaPlus^{TM} QMG 220 (Pfeiffer Vacuum) with a mass range up to 200 amu was used as mass spectrom-

2.7 UV/VIS spectroscopy using an integrating sphere

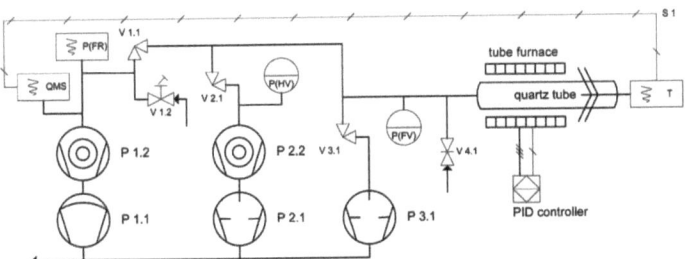

Figure 2.8: Schematic figure of the vacuum system of the TPP-MS system at the Atmospheric Chemistry Research Laboratory of the University of Bayreuth

eter. The filament of the high-sensitivity ion source made of yttriated iridium was operated at 70 eV. A C-SEM (channeltron), operating at 932 V was used as detector. The vacuum system of the sample compartment consisted of a silica quartz tube, located inside the tube furnace, and a vacuum pumping system. The sample was evacuated to fine vacuum using a rotary vane pump P3.1. Afterwards, high vacuum of about 3×10^{-6} mbar was applied using a turbo molecular pump (P2.2) coupled to a rotary vane pump (P.2.1). The vacuum was monitored using a fine-vacuum gauge (P(FV)) and a cold-cathode high-vacuum gauge (P(HV)). For recording the temperature-programmed pyrolysis mass spectra, the samples were heated up to 1000 °C at a constant heating rate of 10 Kmin^{-1}. The furnace was operated using a PID controller (Eurotherm 2216). The actual temperature of the sample was monitored using a Type-K thermocouple. The signal of the thermocouple was transformed to a 0–10 V signal using an Eurotherm 32h8i temperature-monitoring device and was recorded by the mass spectrometer software. All TPP-MS spectra were recorded using the 3-D scan analog setup of the QUADERA software package sampling the full mass range as a function of temperature. Additionally, the total pressure and the sample temperature were recorded.

2.7 UV/VIS spectroscopy using an integrating sphere

The measurement of optical properties of airborne organic aerosols is a complex task. This is especially true for the complex index of refraction, which is the most significant parameter to evaluate optical properties associated with radiative forcing. Hence, only few publications are available on this topic. Myhre et al. (2004) used specular reflectance of liquids containing dissolved organic aerosol acids and a Kramers-Kronig analysis method to obtain the complex index of refraction. Since the method of dissolving aerosols in a liquid provides the complex index of refraction of the solution, the comparability to aerosols in their natural state is questionable. A more comparable approach is the diffuse reflectance measurement of powder samples on filters,

2 Methods

and calculating the complex index of reflection by comparing to transmission measurements (Burger et al., 1997) using the Kubelka-Munk theory (Kubelka and Munk, 1931), see equation 2.5. This technique has also been applied by Campbell et al. (2011). They used a laser to measure the transmission and reflection spectra of aerosols on Teflon filters, hence absorption coefficients could only be derived for the laser wavelength employed. While laser methods only give access to discrete absorption Ångström coefficients, a full-range measurement would be desirable (Moosmüller et al., 2011).

The absolute diffuse reflectivity R_∞ of a sample is proportional to the ratio of the absorption K and the scattering coefficient S (equation 2.5):

$$F(R_\infty) = \frac{(1-R_\infty)^2}{2R_\infty} = \frac{K}{S} \qquad (2.5)$$

The resulting Kubelka-Munk function $F(R_\infty)$ is equal to the extinction and can be related to the Lambert-Beer law. Therefore, a quantitative interpretation of the diffuse reflectance spectra is possible (Kortüm et al., 1963), and the absorption and scattering coefficients can be derived. This is only hampered by a number of limitations which are reported in detail by Kortüm and Oelkrug (1966). One major limitation is the requirement of an infinitely thick sample. According to Kortüm and Oelkrug (1966), a thickness of a few millimeters is required to fulfill the preconditions in the case of very fine powders or particles like atmospheric organic aerosols. It is hardly possible to sample enough aerosol particles to achieve a layer of that thickness, especially not within the available volume of an aerosol smog chamber. Hence, only $F(R)$ and not $F(R_\infty)$ could be measured for the organic aerosols in question.

The best method to determine the complex index of refraction is to measure the scattering of polarized light at different angles (Zhao et al., 1997; Eiden, 1971). However, this sophisticated setup was not available–therefore only the diffuse absorbance of the aerosol samples was recorded.

The UV/VIS diffuse-reflectance spectra were measured using a spectrophotometer (Uvikon XL, BIO-TEK Instruments) with an integrating Ulbricht sphere (Labsphere) between 200 and 800 nm at a speed of 200 nm per minute at a bandwidth of 4 nm. The particulate matter of the aerosol was collected onto Whatman QMA quartz fibre filters (25 mm). Spectralon Diffuse Reflectance Standards (SRS-99-010, Labsphere) were used as reference material for the integrating sphere. No absorption in the chosen spectral range was observed when comparing clean quartz fibre filters to the Labsphere standards using the integrating sphere.

2.8 Ultra-high-resolution mass spectroscopy

To obtain the O/C and H/C ratios as well as the exact amount and the structure of halogenated molecules, ultra-high-resolution mass spectroscopy was performed by Ph. Schmitt-Kopplin at the Helmholtz Centrum Munich (Germany) with a Bruker 12 Tesla APEX Q ion-cyclotron-resonance Fourier-transform mass spectrometer.

Electrospray injection was applied in negative mode with an APOLLO II electrospray source in flow injection at 2 µlmin^{-1} (Gaspar et al., 2009). The molecular formulae were batch-calculated by an in-house software tool, achieving a maximum mass error of ≤ 0.2 ppm. The generated formulae were validated by setting sensible chemical constraints (N rule, O/C ratio ≤ 1, H/C ratio $\leq 2n+2$, element counts: C ≤ 80, H unlimited, O ≤ 60) in conjunction with an automated isotope pattern comparison (Gaspar et al., 2009). Even enabling up to 3 atoms of nitrogen and sulfur, calculated molecules resulted in mainly CHO-types of elemental formulas (only a few formulas corresponding to impurities were found to be CHNO, CHOS, and CHNOS, which originated from the quartz filters).

The filters were pushed into Eppendorf 2-ml vials with the help of the vial cap, extracted directly with 1 ml of methanol, and centrifuged in the same Eppendorf vials. The extract was used without further treatment.

2.9 Sample preparation techniques

The SOA precursors were introduced into the aerosol smog chambers using an impinger connected to the aerosol smog chamber and flushed with purified air. In case of liquid precursors, a calculated amount of the sample, based on the chamber volume, was injected into the impinger using a µL-syringe and flushed into the chamber. While the vapor pressure of α-pinene is high enough (500 Pa) to evaporate the precursor only by the air flow, the guaiacol precursor (vapor pressure of 14.66 Pa) had to be heated to facilitate fast evaporation. The required amount of the solid SOA precursor catechol (vapor pressure of 20 Pa) was weighed, inserted into the impinger, heated to liquefaction, and then flushed into the aerosol smog chamber like the other precursors.

For aerosol flow-reactor experiments, the impingers were permanently filled with the precursors and flushed. In the case of the catechol precursor, the impinger was permanently heated to liquefy the solid compound (110 - 120 °C).

2 Methods

For the molecular processing experiments, gaseous reactants like chlorine and bromine were also introduced through the impinger. While chlorine was diluted with purified air in a ratio of 1:100 and then dosed using a mL-syringe, bromine was dosed as a liquid with a µL-syringe.

To generate the sea-salt aerosol for the sea-salt SOA interaction experiments. Sea-salt aerosol was added to the aerosol smog chamber with the already formed SOA inside. To add the sea-salt aerosol without diluting the SOA in the smog chamber, an internal nebulizer was built, based on the concept reported by Siekmann (2008). The internal nebulizer uses air from the smog chamber to nebulize the sea-salt aerosol into the chamber (Figure 2.9).

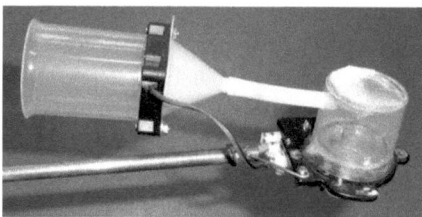

Figure 2.9: Internal nebulizer to add sea-salt aerosol to an existing aerosol in the aerosol smog chamber (developed by H.-U. Krüger)

3 SOA formation

A detailed characterization of the SOA model employed is given in this chapter. As introduced in chapter 1, α-pinene, catechol and guaiacol were used as precursors to form the SOA by homogeneous nucleation in the aerosol smog chamber. Since SOA from α-pinene is a well-characterized organic aerosol, SOA from α-pinene was only investigated (see section 3.2) with respect to the characterization of changes induced by the halogen-aerosol interaction. Due to the fact that SOA from catechol or guaiacol is rather unknown, their nucleation products were characterized in more detail (see section 3.3).

3.1 Experimental setup and materials

For aerosol smog-chamber experiments, the following SOA precursors were used. As organic precursor for the aliphatic SOA, α-pinene was obtained (Figure 1.2) from Sigma-Aldrich with a purity > 98 %. The aromatic precursors for the SOA were catechol (Riedel-de Haën, 32101, pro-analysis grade, > 99 % HPLC) and guaiacol (Sigma Aldrich, G5502, pro-analysis grade, > 99 % GC) (Figure 1.4). The molecular weights are: α-pinene, 136.24 g mol^{-1}; catechol, 110.11 g mol^{-1}; guaiacol, 124.14 g mol^{-1}.

For aerosol flow-reactor experiments, the same α-pinene precursor was used. The precursor for the catechol aerosol was obtained from Merck with a purity > 99 % by GC area.

The precursors and their physicochemical properties are listed in table 3.1. The sample preparation technique for the aerosol smog chamber and the aerosol flow reactor is described in detail in section 2.9. The ozone production and the purity of the oxygen used was described in detail above in section 2.1.

For each precursor three aerosol smog-chamber experiments were carried out: 1. formation of SOA in the dark with O_3 only (0 % relative humidity); 2. formation of SOA with simulated sunlight and O_3 only (0 % relative humidity); 3. formation of SOA with simulated sunlight, O_3 and 25 % relative humidity. No particle formation could be observed from the reaction of guaiacol with O_3 in the dark, therefore, no data is presented. For aerosol size distribution experiments, 100 ppb of the precursor and 500 ppb of ozone were used to record the formation

3 SOA formation

Table 3.1: SOA precursors and their significant physicochemical properties

precursor	molecular weight [g mol^{-1}]	density [g cm^{-3}]	vapor pressure [Pa]
α-pinene	136.24	0.858	500
catechol	110.11	1.34	20
guaiacol	124.13	1.128	14.66

of the SOA. FTIR experiments had to be performed at higher concentrations (5 ppm precursor and 20 ppm ozone) due to the detection limit of the long-path FTIR spectrometer. Two hours after SOA formation, ATR samples were collected onto the KRS-5 crystal for 30 minutes at an aerosol flow of 6.5 cm^{3}s^{-1}. About 10^8 aerosol particles were collected for each ATR measurement from medium particle concentrations of 10^5 *particles* cm^{-3} at a collection efficiency of about 90 %. For UV/VIS spectroscopy, FEG-SEM imaging, TPP-MS, and ultra-high-resolution mass spectroscopy, aerosol particles formed at higher precursor concentrations (5 ppm) were sampled onto the specified filter materials.

Aerosol flow-reactor experiments were performed at higher precursor and ozone concentrations (α-pinene: 5000 ppm, catechol: 50 ppm; ozone up to 500 ppm). All secondary organic aerosols were produced only by ozone oxidation without UV irradiation.

3.2 SOA from α-pinene

Because there is already a huge amount of journal articles on SOA formation from the terpene-type precursor α-pinene available, only spectroscopic details were investigated within this work to be able to determine physicochemical changes of the SOA through the reaction with halogen species.

3.2.1 Particle formation in the 700 L aerosol smog chamber

Various groups have studied the SOA formation from terpene-type precursors, e.g. Baltensperger et al. (2005). The particle formation shown in figure 3.1 (the so-called "banana plot"), which was measured in the 700 L aerosol smog chamber, demonstrates the rapid SOA formation caused by the high precursor concentrations needed for applying spectroscopic techniques, as described in chapter 2.

Caused by the rather high precursor concentrations, the progression of the homogeneous nucleation is barely observable. The precursor and oxidant concentrations mentioned above

3.2 SOA from α-pinene

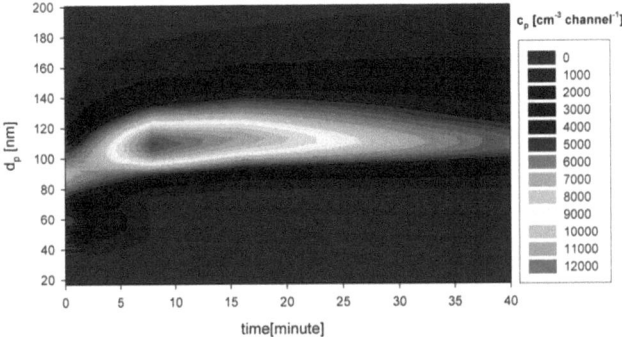

Figure 3.1: Formation of SOA from α-pinene in the 700 L aerosol smog chamber

lead to SOA particles with diameters in the range of 100–120 nm. The aerosol formation process is completed rather quickly. After 10 minutes, no further significant changes are visible except a decrease in particle concentration caused by sedimentation and diffusion to the chamber walls.

3.2.2 Infrared spectroscopy of particle formation in the smog chamber

To characterize SOA from α-pinene formed in the 700 L aerosol smog chamber in detail, long-path absorption FTIR spectroscopy was used. The formation of this terpene-type SOA is very fast, hence only a few spectral changes during a smog-chamber run are visible in the infrared spectrum (Figure 3.2.

The time-resolved infrared spectra are dominated by the formation of a broad $\nu(O-H)$ stretch region, a strong $\nu(C=O)$ absorption, and the degradation of the $\nu(C-H)$ region. In comparison with the infrared spectra of the precursor (Wilson, 1976) and with an other time-resolved FTIR study (Sax et al., 2005), the absorptions occurring in figure 3.2 could be assigned as follows.

The high-wavenumber range is dominated by a broad $\nu(O-H)$ vibration with a maximum at about 3420 cm^{-1}. According to Sax et al. (2005), these stretching vibrations can be assigned to aliphatic alcohols. The stretching region of the $-OH$ groups of carboxylic acids, between 3200 and 2400 cm^{-1}, is weakly structured in the long-path absorption spectrum and therefore discussed elsewhere (see section 3.2.4, ATR spectra of SOA from α-pinene).

The $\nu(C-H)$ region is dominated by the asymmetric and symmetric stretching vibrations of the $-CH_3$ and $-CH_2-$ groups. The strongest absorption at 2924 cm^{-1} correlates with the $\nu_{as}(C-H)$ of a $-CH_2-$ group. The intensity of this absorption decreases faster than that of the

3 SOA formation

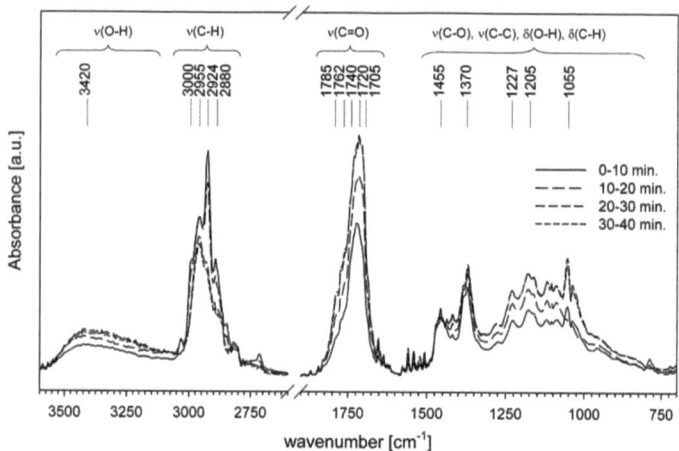

Figure 3.2: Time-resolved long-path infrared absorption spectra of SOA formation from α-pinene with ozone and without simulated sunlight

nearby absorption at 2955 cm^{-1}, which is the $\nu_{as}(C-H)$ of a $-CH_3$ group. At the beginning of the aerosol formation process, the asymmetric stretching vibration of $-CH_2-$ groups at 2924 cm^{-1} is more prominent than the asymmetric stretching vibration of the $-CH_3$ groups at 2955 cm^{-1} (Wilson, 1976). This is in good agreement with ring-opening reactions and degradation of aliphatic $-CH_2-$ groups (Yu et al., 2008).

Thus, an oxidation of this aliphatic group plays an important role for SOA formation from α-pinene. Further prominent $\nu(C-H)$ vibrations are a second $\nu_{as}(C-H)$ at about 3000 cm^{-1} and a $\nu_s(C-H)$ at 2880 cm^{-1}. Other, less important vibrations in this spectral range can be assigned in good agreement with Wilson (1976).

The absorption intensity within the carbonyl stretching region between 1800 and 1650 cm^{-1} is increasing due to the oxidation of the precursor resulting in the formation of $C=O$-containing functional groups. The main absorption can be found at about 1720 cm^{-1} (reported by Sax et al. (2005) as 1726 cm^{-1}). In total, contributing to the carbonyl stretching region five absorptions could be identified: 1785, 1762, 1740, 1720, and 1705 cm^{-1}. A definite assignment of these absorptions to functional groups is not yet possible, but in general the following dependencies were considered: Unsaturated aldehydes and acids can be found at 1720 and 1705 cm^{-1}. Saturated ketones, aldehydes, and esters contribute to the absorption at 1740 cm^{-1}. Carboxylic acid anhydrides and stable aliphatic acid peroxides could be found at 1762 and 1785 cm^{-1}. The allocation of carbonyl-containing functional groups was based on the detailed description of the

3.2 SOA from α-pinene

Table 3.2: Assignment of infrared absorptions of SOA from α-pinene in the smog chamber

vibr. mode	[cm^{-1}]	assignment
$\nu(O-H)$	3420	mainly different alcohols
$\nu_{as}(C-H)$	3000 & 2955	$-CH_3$
$\nu_{as}(C-H)$	2924	$-CH_2-$
$\nu_s(C-H)$	2880	$-CH_3$
$\nu(C=O)$	1785 & 1762	acids, anhydrides, peroxides
$\nu(C=O)$	1740	saturated ketones, aldehydes, esters
$\nu(C=O)$	1720 & 1705	unsaturated acids, aldehydes
	1455	-"-
$\delta(C-H)$	1370	C-H in-plane deformation
	1227	-"-
$\omega(C-H)$	1205	wagging of $-CH_2$ and $-CH_3$
$\nu(C-C)$ and/or $\alpha(ring)$	1055	skeleton vibrations

carbonyl stretching region by Socrates (1980). This region was further investigated using the aerosol flow reactor (see section 3.2.3).

The following fingerprint region is dominated by residual absorptions of the entire precursor. The band at 1445 cm^{-1} has been observed by Wilson (1976), but was reported without detailed vibrational assignment. The $C-H$ in-plane deformation vibration is visible at 1370 cm^{-1}. No detailed information on the absorption at 1227 cm^{-1} is available. At 1205 cm^{-1}, the wagging vibration of aliphatic $-CH_x$ is visible. The lowest prominent absorption, at 1055 cm^{-1}, represents carbon-skeleton vibrations of the ring system, but it might be concealed by the ozone vibration at 1043 cm^{-1}.

The assignments of the most prominent absorptions of the long-path FTIR spectra are listed in table 3.2.

3.2.3 Spectroscopy of particle formation using the aerosol flow reactor

The aerosol flow reactor was used to study the formation of SOA from α-pinene with better time resolution. The absorbance was observed to increase over time (Figure 3.3 AFR), which is due to the ongoing deposition of organic material onto the KBr windows of the gas cell. The resulting infrared spectra are negligibly influenced by this contamination of the windows, because the signal of the gas phase is significantly stronger.

The results from the aerosol flow reactor are confirmed by the change in predominant CH_x-groups. These vibrations show a quite similar behavior in the long-path FTIR spectra. In

3 SOA formation

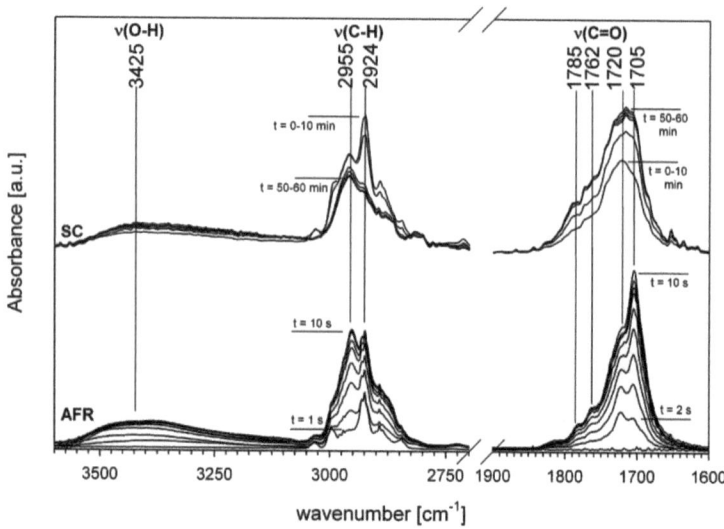

Figure 3.3: Formation of SOA from α-pinene studied inside an aerosol smog chamber (SC, temporal resolution 10 min) and an aerosol flow reactor (AFR, temporal resolution 1 s up to 10 s)

the course of the measurement, the most outstanding asymmetric stretching vibration of the $-CH_2-$ group at 2924 cm^{-1} is replaced by the asymmetric stretching vibration of the $-CH_3$ group at 2955 cm^{-1}. In contrast to the temporally averaged spectra of the White cell, many bands of the entire core structure of the α-pinene molecule remain visible during the first few seconds. Studying the formation of SOA from α-pinene using an aerosol flow reactor provides a more detailed understanding of the formation of the $\nu(C=O)$ stretching vibration (Figure 3.3 AFR). The AFR spectra indicate a composition of the broad carbonyl vibration of two main absorptions at 1720 and 1705 cm^{-1}. At higher frequencies, shoulders at 1785 and 1762 cm^{-1} are also visible. At the beginning of the reaction of ozone with α-pinene, the absorption at 1720 cm^{-1} is predominating. It can be assigned to the $\nu(C=O)$ stretching vibrations of saturated ketones or aldehydes (Socrates, 1980). Later on, the band at 1705 cm^{-1} becomes more intensive. This band represents saturated carboxylic acids or α-diketones. Both absorptions together might be assigned to saturated dicarboxylic acids, but in this case the observed switch in relative intensity should not occur. The two bands at the high frequency shoulder can be assigned to carboxylic acid anhydrides, which is in good agreement with the rise in intensity of the $\nu(C=O)$ stretching vibration of carboxylic acids at 1705 cm^{-1}.

3.2 SOA from α-pinene

Table 3.3: Assignment of infrared absorptions by SOA from α-pinene using the aerosol flow reactor

vibr. mode	[cm^{-1}]	assignment
$\nu_{as}(C-H)$	2955	$-CH_3$
$\nu_{as}(C-H)$	2924	$-CH_2-$
$\nu(C=O)$	1785 & 1762	carboxylic acid anhydrides
$\nu(C=O)$	1720	saturated ketones and aldehydes
$\nu(C=O)$	1705	unsaturated acids or α-diketones

A list of all examined absorptions using the aerosol flow reactor is given in table 3.3. A quantitative analysis of the time-dependent increase or decrease of absorptions is hardly possible because of the poor time resolution of the aerosol smog chamber and the above-mentioned contamination of the windows of the absorption cell of the aerosol flow reactor.

3.2.4 Spectroscopy of functional groups of particulate matter

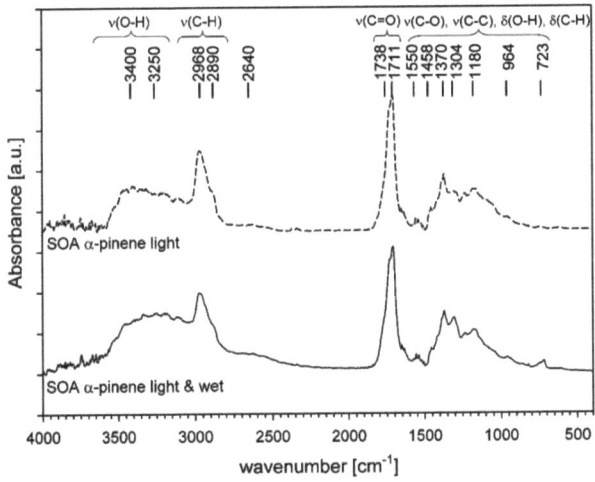

Figure 3.4: ATR-FTIR spectra of particulate matter of SOA from α-pinene at two different ambient conditions

The particulate matter of SOA from α-pinene was analyzed separately using ATR-FTIR spectroscopy. The SOA was formed at two different ambient conditions: with simulated sunlight

3 SOA formation

Table 3.4: Assignment of ATR infrared absorptions from SOA formation from α-pinene in the smog chamber

vibr. mode	[cm^{-1}]	assignment
$\nu(O-H)$	3400	mainly different alcohols
$\nu(O-H)$	3250 & 2640	carboxylic acids
$\nu_{as}(C-H)$	2968	$-CH_3$
$\nu_s(C-H)$	2890	$-CH_3$
$\nu(C=O)$	1738	saturated ketones, aldehydes, esters
$\nu(C=O)$	1711	unsaturated acids, aldehydes
$\nu(C=C)$	1550	$-CH=CH-$ of the precursor ring
	1458	-"-
$\delta(C-H)$	1370	in-plane deformation
$\nu(C-O)$	1304	carboxylic acid dimers and/or esters
$\nu(C-H)$	1180	in-plane deformation
$\delta(C-H)$	723	out-of-plane, cis $-CH=CH-$

at 0 % relative humidity and at 25 % relative humidity. Sampling onto the ATR crystals was performed using electrostatic precipitation (see section 2.5.3).

The $\nu(O-H)$ region of the ATR spectra of the two different samples is dominated by a broad absorption with its maximum located at about 3400 cm^{-1} for SOA from α-pinene formed with simulated sunlight at 0 % relative humidity and at about 3250 cm^{-1} for SOA from α-pinene formed with simulated sunlight at 25 % relative humidity. While the higher-wavenumber region represents alcohols and intramolecular hydrogen bonds of -OH groups, the absorption at 3250 cm^{-1} can be related to the band at 2640 cm^{-1}. Those two vibrational regions are typical of $\nu(O-H)$ of carboxylic acids.

At 2968 and 2890 cm^{-1}, two dominant absorptions of $-CH_x$ groups are present in the SOA from α-pinene. Confirmed by the assignments for the precursor, 2968 cm^{-1} is the asymmetric and 2890 cm^{-1} the symmetric stretching vibration of the $-CH_3$ groups.

In addition to the carbonyl stretching region described in sections 3.2.2 and 3.2.3, two main absorptions are visible at 1738 and 1711 cm^{-1}. Those can be assigned the same way as in sections 3.2.2 and 3.2.3.

Apart from the $O-H$ stretching region, significant differences between the different ambient conditions applied are only visible in the so-called fingerprint region. While absorptions like 1550, 1438, and 1370 cm^{-1} seem to be rather similar for both samples, the vibrations at 1304 and 723 cm^{-1} are more prominent at enhanced relative humidity. Only a few vibrations can be assigned to structural elements of the precursor: The presence of the ring system of the

3.2 SOA from α-pinene

precursor is revealed by the in-plane deformations at 1370 and 1180 cm^{-1} and the $\nu(C=C)$ at 1550 cm^{-1}. A strong absorption close to 1458 cm^{-1} was measured for the precursor (Wilson, 1976) but was not assigned to any specific vibration of the precursor molecule. The same is true for the weak absorption at about 1550 cm^{-1}. Strong hints at carboxylic acid dimers or esters are given by the increasing $\nu(C-O)$ vibration at 1304 cm^{-1} in combination with the $\nu(O-H)$ region. A deformation vibration of the unoxidized $-CH=CH-$ is present in the sample formed at enhanced relative humidity.

While the ATR-FTIR spectra of SOA from α-pinene exhibit more spectral features than described above, this section only focuses on the main absorptions as listed in table 3.4.

The method of TPP-MS to study functional groups of organic macromolecular structures is useful for low-volatile materials like soot (Muckenhuber and Grothe, 2004; Ofner and Grothe, 2007) or low-volatile organic aerosols (see section 3.3). The important TPP-MS mass peaks of SOA from α-pinene appear at the very beginning of the desorption process during the evaporation of semi-volatile compounds. Hence, an assignment to functional groups as done by Muckenhuber and Grothe (2004) is hardly possible. The complete TPP mass spectrum is shown in figure 3.5.

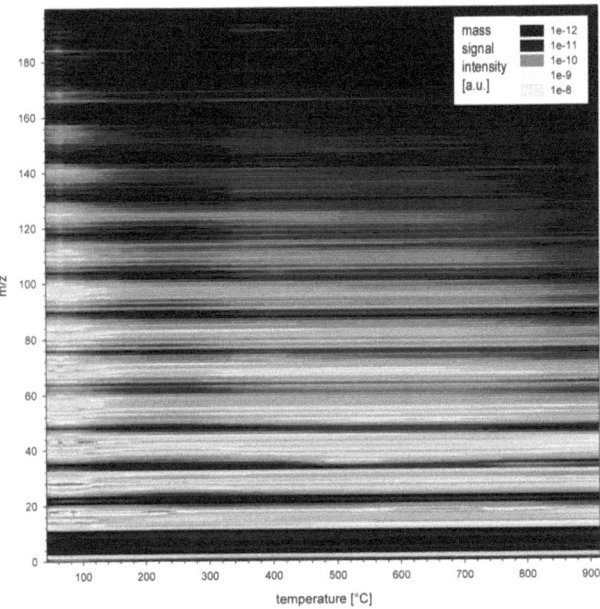

Figure 3.5: TPP mass spectra of particulate matter of SOA from α-pinene

3 SOA formation

The semi-volatility of SOA from α-pinene is demonstrated by the high peaks over the whole mass range during desorption below 100 °C. However, also low-volatile compounds are part of the particulate matter, which seems to be stable up to 900 °C. Some hints of oligomer formation are given by the separated desorption blocks, which have an average width of about 12 m/z. In TPP mass spectra, only compounds not exceeding 130 m/z are stable up to high temperatures. This might be caused by fragmentation of oligomers with higher masses. This fragmentation could be due to a rather weak bonding of the oligomers, e.g. by carboxylic acid dimers or hydrogen bonding as indicated by the ATR spectra.

An assignment of CO, CO_2, and OH desorption signals to functional groups according to Muckenhuber and Grothe (2006) is impossible, because a strong desorption of those fragments is even visible in the range between room temperature and 100 °C. This behavior indicates that the vapor pressure of some compounds of the α-pinene SOA is not low enough to permit an analysis at high-vacuum conditions.

3.2.5 Optical properties

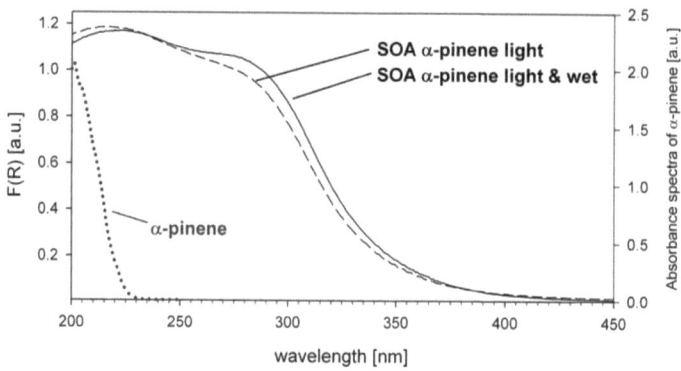

Figure 3.6: Diffuse reflectance spectra $(F(R))$ of SOA from α-pinene and absorption spectra of a saturated α-pinene vapor

The diffuse-reflectance UV/VIS spectra of SOA from α-pinene formed with simulated sunlight and 0 % or 25 % relative humidity exhibit almost no absorption in the visible part of the spectrum, i.e. above 400 nm (Figure 3.6), which is in good agreement with the nearly colourless appearance of the filter samples. The absorption spectrum of saturated α-pinene vapor was calculated based on the absorption cross section (Kubala et al., 2009), published in the MPI-Mainz UV-VIS spectral atlas of gaseous molecules.

The main absorptions of SOA from α-pinene are located at about 220 and 275 nm. They can be related to the $\pi \to \pi^*$ and the $n \to \pi^*$ transition. The absorption at 275 nm is due to oxygen-containing functional groups containing lone pairs of electrons, like carboxylic acids and carbonyls. The absorption at 220 nm appears to originate from short conjugated systems, which stem from the $-CH=CH-$ bond of the α-pinene ring system (Hesse et al., 1991). These might be e.g. conjugated dienes. All spectra shown in figure 3.6 exhibit a nearly exponential decay towards the red end of the spectrum.

3.3 SOA from catechol and guaiacol

3.3.1 Particle formation - particle number concentration and size distributions

The aerosol size distributions of SOA from catechol and guaiacol were measured as described in section 2.3. Two aerosol size distributions are compared: one is of SOA formed in the presence of simulated sunlight and 25 % relative humidity and the other of SOA formed in the dark at 0 % relative humidity (Figure 3.7). Particle formation is significantly enhanced by UV/VIS irradiation and enhanced relative humidity.

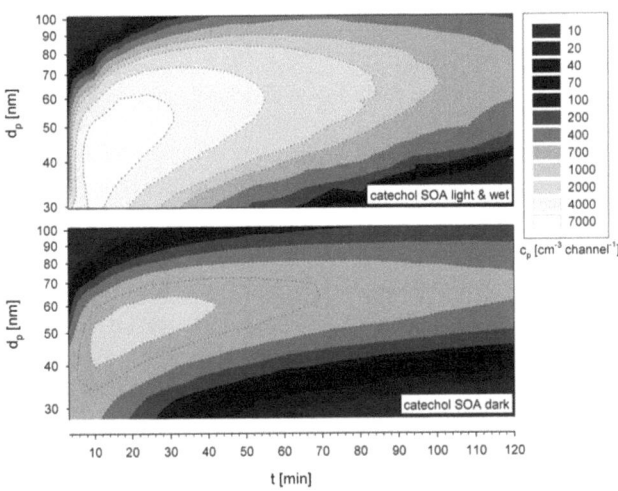

Figure 3.7: Aerosol size distribution of SOA from catechol formed in the presence of simulated sunlight and 25 % relative humidity and formed in the dark at 0 % relative humidity

51

3 SOA formation

The mass concentrations and size distributions of the SOA formed strongly depend on the environmental conditions in the aerosol smog chamber, as indicated in figure 3.7. More detailed information on the particle formation process and the achieved masses is given in figure 3.8, where the medium particle diameters of the different aerosols are plotted against their formation time. The aerosol mass distributions were derived from volume distributions with an assumed density of 1.4 g cm^{-3} (Coeur-Tourneur et al., 2009).

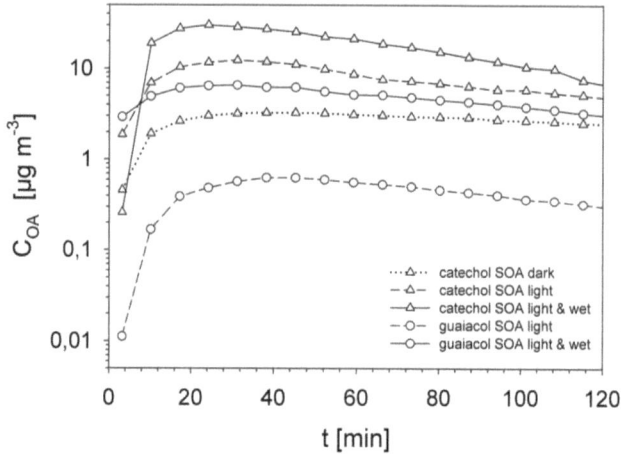

Figure 3.8: Comparison of the evolution of the aerosol mass concentration (C_{OA}) at different simulated environmental conditions and depending on the SOA precursors.

Aerosol formation yields are influenced by the presence of simulated sunlight and relative humidity. The precursors play a decisive role in this formation process. The catechol precursor is found to generate one order of magnitude more aerosol mass than the guaiacol precursor under simulated-sunlight conditions. At a relative humidity of 25 %, these yields are observed to increase significantly compared to the yields at dry conditions or without simulated sunlight. SOA formation experiments from catechol and guaiacol (formed at 25 % relative humidity), particle size distributions show similar mean particle diameters of 50 ± 5 nm. The width of the particle size distribution for SOA formed in the dark (σ_g=1.26) is lower than that for SOA formed under simulated-sunlight conditions (σ_g=1.34). The mean particle diameter of SOA from guaiacol formed in the presence of simulated sunlight and dry conditions is very low (32 nm). This is caused by the reduced reactivity of guaiacol towards ozone.

Thus, irradiation has an influence on particle formation, not only concerning the overall aerosol mass, but also the particle size distribution as shown in figure 3.7. Particle formation

3.3 SOA from catechol and guaiacol

seems to be fairly completed after about 30 minutes. Subsequently, only aggregation of small particles takes place, as demonstrated by the slow increase in aerosol diameters. A detailed discussion of the resulting morphology, based on electron microscopy, is given in the next section (3.3.2).

3.3.2 Particle imaging using FEG-SEM

To study the morphology of SOA formed from catechol and guaiacol, FEG-SEM (see section 2.4) was used. Aerosol particles were deposited either on Si wafers using electrostatic precipitation (for details see section 2.5.3), or collected using polycarbonate filters as described in section 2.4. For scanning electron microscopy, higher precursor concentrations were employed to achieve sufficient SOA particle density. As a result, aerosol particles were slightly larger than the mean value observed in the aerosol size distributions (Section 3.3.1).

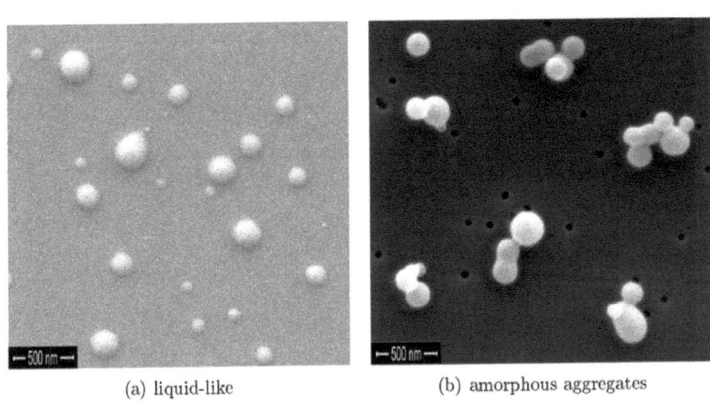

(a) liquid-like (b) amorphous aggregates

Figure 3.9: FEG-SEM images of SOA from catechol, exhibiting two different morphologies

Figure 3.9 displays aerosol particles on a Si wafer (3.9(a)), which exhibit a rather liquid-like morphology, and particles on a polycarbonate filter (3.9(b)), most of which are aggregates with an amorphous structure. The morphological difference between those two samples could be caused either by the different sampling methods or by different states of aging of the aerosols. Due to the fact that the method of electrostatic precipitation was verified with regard to chemical transformations of the particulate matter (Ofner et al., 2009), different states of oxidation of the aerosols are more likely.

Assuming that the liquid-like particles take on a hemispherical shape when deposited on the wafer, the diameters of the airborne particles are 1.26 times smaller than the diameters

3 SOA formation

suggested by the image. This is caused by the different diameters of particles of the same volume in spherical and hemispherical shape.

The amorphous particles in figure 3.9(b), with diameters between 100 and 250 nm, are of nearly perfect spherical morphology, indicating their formation by gas-to-particle conversion while being airborne (Pöschl et al., 2010). Du no specific surface texture is visible. Chain- and cluster-like aggregates of those particles have also been observed.

3.3.3 Infrared spectroscopy of particle formation in the smog chamber

Long-path absorption infrared spectroscopy of the SOA formation from catechol and guaiacol in the aerosol smog chambers allows to follow both, the decrease of the precursors and the formation of the organic aerosol. The mean residence time τ of catechol in the smog chamber was calculated to be 10 minutes by observing the decrease of the aromatic stretching vibration at 1510 cm^{-1}. Hence, during the first 10 minutes the precursor decreases rapidly and products increase with approximately the same rate.

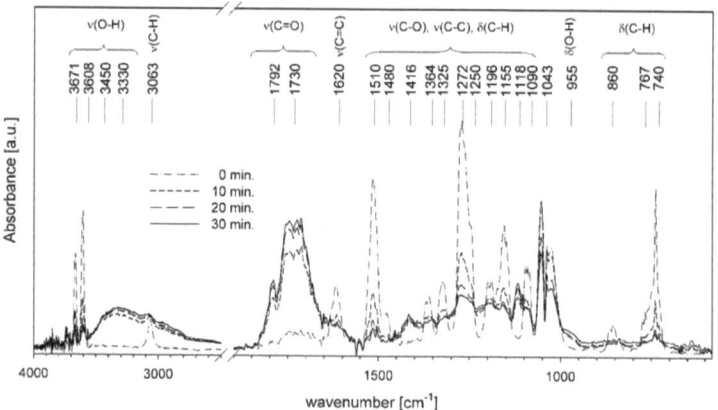

Figure 3.10: Time-resolved long-path infrared absorption spectra of SOA formation from catechol with ozone and without simulated sunlight: The precursor (dash-dotted) rapidly decreases and the broad, largely unstructured bands of the organic aerosol (black) increase at the same time.

After 30 minutes, no further changes in the infrared spectrum were detectable under the given experimental conditions. Thus, the formation process of SOA particles from catechol or guaiacol as precursors is completed after less than 30 minutes (Figure 3.10). The long-path FTIR spectra from the aerosol smog chamber indicate that the fully developed particles do

3.3 SOA from catechol and guaiacol

not undergo any further oxidation. This can be easily inspected by the carbonyl stretching vibration (see figure 3.10), which then remains constant in shape and intensity. Therefore, even the chemical aging of the organic aerosol seems to be finished after 30 minutes, and the state of oxidation is stable under the given experimental conditions.

Corresponding to the composition of the gas phase (N_2, O_2, CO_2, and O_3) and the chemical structure of the precursors, only vibrational modes containing carbon, oxygen, and hydrogen are expected. The overlap of absorptions of gaseous and solid products as well as gaseous educts and other reactive or inert gas-phase species constrains the quantitative analysis of these data.

At the beginning of the SOA formation, the spectra of the organic gas or particle phase can be assigned to the catechol molecule (Socrates, 1980). The aromatic ring exhibits typical aromatic $\nu(C=C)$ stretching vibrations at 1620, 1510, and 1480 cm^{-1}. The sharp absorption at 3063 cm^{-1} matches the $\nu(C-H)$ stretching vibration, the two absorptions at 3671 and 3608 cm^{-1} are the respective $\nu(O-H)$. The vibrations at 1364 and 1325 cm^{-1} correspond to the phenolic $\nu(C-O)$ stretching vibration of phenolic alcohols. The absorptions in the range of 1272 to 1155 cm^{-1} and the band at 1090 cm^{-1} represent either the aromatic in-plane deformation $\delta(C-H)$ or the phenolic deformation mode $\delta(O-H)$. The aromatic out-of-plane deformations give rise to the absorptions between 860 and 740 cm^{-1}. The ozone content is indicated by the absorption at 1043 cm^{-1}.

During the aerosol formation process most of those absorptions decrease, and the formation of some new bands can be observed. The sharp maxima of the phenolic $\nu(O-H)$ vibration decrease, and new $\nu(O-H)$ vibrations emerge, as indicated by the broad absorption at about 3330 cm^{-1}. On the high-frequency slope of this group of bands, the absorption at 3450 cm^{-1} increases as well. This band might be assigned to the intra-molecularly bonded $\nu(O-H)$ of $-O \cdots H \cdots O =$.

A strong carbonyl stretching vibration in the range of 1850 to 1680 cm^{-1}, while the aerosol is formed. Two main bands can be identified in this region using long-path absorption FTIR in the aerosol smog chamber. An absorption at 1792 cm^{-1} indicates the $\nu(C=O)$ of esters, anhydrides, and carboxylic acids, and an absorption at 1730 cm^{-1} is typical of quinones, ketones, and other aromatic and aliphatic $\nu(C=O)$ vibrations. A more detailed characterization of the carbonyl stretching region is given in the following section 3.3.4.

The increasing vibration at 1416 cm^{-1} is interpreted as the $\delta(O-H)$ deformation vibration in combination with the $\nu(C-O)$ stretching vibration of carboxylic acids or phenols. Strong hints at aliphatic or aromatic ether formation are given by the vibration at 1118 cm^{-1}. This vibration might represent the aliphatic or aromatic $\nu(C-O)$ stretching vibration of ethers. The presence of carboxylic acids is underlined by the appearance of an absorption at 955 cm^{-1}, indicating the out-of-plane deformation mode $\delta(O-H)$.

3 SOA formation

The decrease of the sharp maxima, which can be assigned to the aromatic ring vibrations, points at partial ring-opening reactions. However, aromatic and unsaturated structures are still present in the resulting aerosol particles, as indicated by the band at 1620 cm^{-1}. Therefore, the resulting organic molecules still contain higher-oxidized benzene fragments.

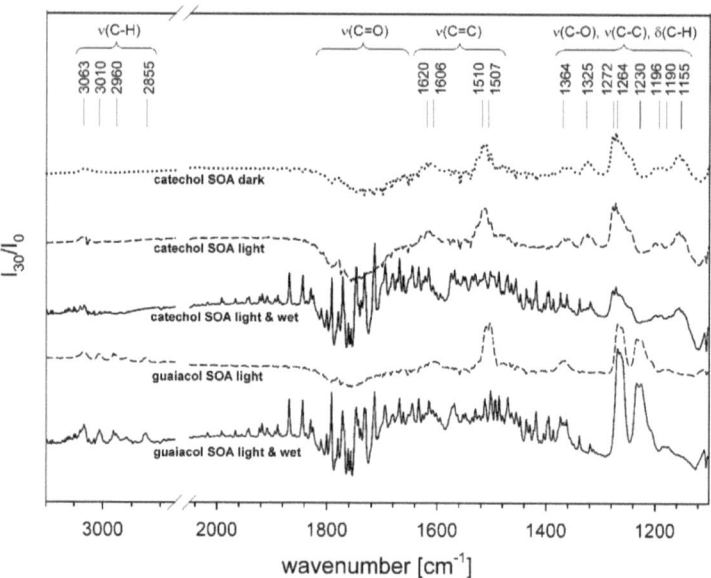

Figure 3.11: Long-path absorption infrared spectra of catechol and guaiacol compared to SOA formed after 30 minutes at different ambient conditions.

The comparison of the infrared transmission spectra of organic aerosols formed at different ambient conditions highlights functional differences not only resulting from different precursors, but also from different experimental conditions (Figure 3.11). The transmission spectra were calculated by dividing the single spectra of the organic aerosols after 30 minutes reaction time by the single spectra of their respective precursors at the beginning of the smog-chamber run.

The $\nu(C-H)$ stretching vibration of the aromatic hydrogens at 3063 cm^{-1} decreases more markedly under wet conditions. This decrease appears to be the strongest with the "wet" guaiacol aerosol. The absorptions at 3010, 2960, and 2855 cm^{-1}, which only appear in the guaiacol aerosol, are assigned to methyl ethers. This group is lost in the course of the aerosol formation process, which is most pronounced under wet conditions. Carbonyl-containing functional groups involving $\nu(C=O)$ absorptions between 1850 and 1690 cm^{-1} cannot be characterized in detail, because they are concealed by the infrared absorption of water. The $\nu(C=C)$ stretching vibration of the aromatic ring at 1620 cm^{-1} for the catechol-derived aerosol is shifted to 1606 cm^{-1} for the guaiacol-derived aerosol because of the higher total mass of the precursor molecule. The same shift occurs for the aromatic stretching vibration at 1510 to 1507 cm^{-1}.

The decrease of the phenolic group absorption at 1364 and 1325 cm^{-1} is visible for all five ambient conditions employed, although the band at 1325 cm^{-1} is not visible in the spectra of guaiacol aerosol. The aromatic ring deformation modes at 1272 and 1155 cm^{-1} also decrease during the conversion of the precursor. Differences within this frequency range agree with the lowered symmetry of guaiacol. The absorptions of all other main structural elements and functional groups also showed signs of formation and degradation processes. Main differences arise from the additional methyl-ether of guaiacol and therefore from the reduced symmetry and the increased mass of the precursor.

The most important observed absorptions indicating the formation or degradation of functional groups and structural elements during SOA formation from catechol and guaiacol are summarized in table 3.5. This table only contains infrared absorptions observed using long-path absorption FTIR spectroscopy.

3 SOA formation

Table 3.5: Assignment of infrared absorptions from SOA from catechol and guaiacol using long-path FTIR.

vibr. mode	[cm^{-1}]	assignment	catechol SOA[1]			guaiacol SOA[1]	
			dark	light	light & wet	light	light & wet
$\nu(O-H)$	3450	intra-molec. & -COOH	m	m	m	m	m
$\nu(O-H)$	3330	different -OH	m	m	m	m	m
$\nu(C-H)$	3063	aromatic =CH	m	w	m	w	w
$\nu(C-H)$	3010	aliphatic $-CH_x$ of guaiacol				m	w
$\nu(C-H)$	2960	aliphatic $-CH_x$ of guaiacol				m	w
$\nu(C-H)$	2855	aliphatic $-CH_x$ of guaiacol				m	w
$\nu(C=O)$	1792	esters, anhydrides & carboxylic acids	s				
$\nu(C=O)$	1730	quinones & ketones	s				
$\nu(C=C)$	[2]	aromatic, olefinic	1620	1620	1620	1606	1606
$\nu(C=C)$	[2]	aromatic ring vibr.	1510	1510	1510	1507	1507
$\delta(O-H)$	1416	of -COOH	s	s	m	m	m
$\nu(C-O)$	1364	degradation of -OH	m	m	m	s	s
$\nu(C-O)$	1325	degradation of -OH	m	m	m		
$\delta(C=C)$	1272	aromatic ring degradation	m	m	m	s	s
$\delta(C=C)$	1155	aromatic ring degradation	m	m	m		
$\nu(C-O)$	1118	of R-O-Ar [a]	m	w	w	w	w
$\delta(O-H)$	955	-COOH	m	w	w	w	w

[a] Ar - representing conjugated or aromatic structural elements; R - representing aliphatic or substituted olefinic structural elements
[1] w = weak, m = medium, s = strong
[2] shift of absorption depending on the precursor

3.3.4 Spectroscopy of particle formation using the aerosol flow reactor

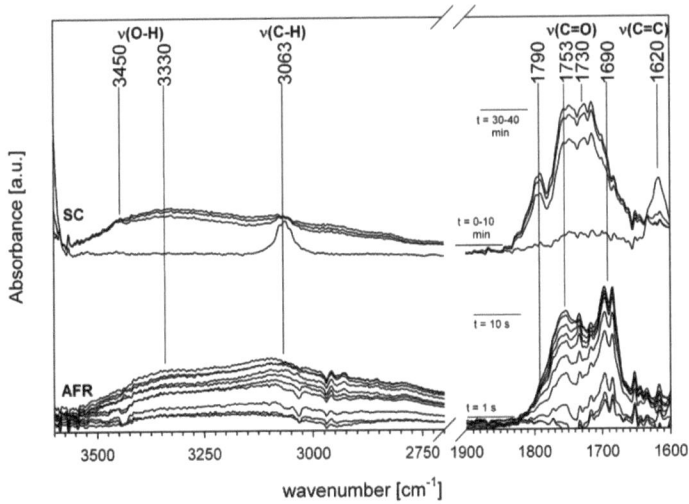

Figure 3.12: Infrared spectra of the formation of SOA from catechol studied in an aerosol smog chamber (SC, temporal resolution 10 min) and an aerosol flow reactor (AFR, temporal resolution 1 s)

Figure 3.12 shows a comparison between the time-dependent aerosol formation process in the aerosol smog chamber and the aerosol flow reactor. The time-resolved infrared spectra of aerosol flow-reactor experiments from catechol exhibit more details in the region of carbonyl vibrations (Figure 3.12). Again, the vibration of carboxylic acid anhydrides at 1790 cm^{-1} is still present, but not as intensive as in the long-path FTIR spectra. The broad absorption at 1730 cm^{-1} resolves into two absorptions at 1753 and 1690 cm^{-1}. The strong absorption at 1690 cm^{-1} can be assigned to aryl ketones, quinones, or aryl carboxylic acids (Socrates, 1980), representing the aromatic structural element of the resulting SOA particles. This absorption may also arise from various compounds based on the structure of trans,trans-muconic acid, for which the carbonyl vibrations are reported at 1695 cm^{-1} (Bejan, 2006). The absorption at 1690 cm^{-1} can also be assigned to o-benzoquinone. The infrared spectrum of p-benzoquinone, which should show only minor differences to the o-benzoquinone spectrum, is reported in the NIST/EPA gas-phase infrared database. The weaker absorption at 1753 cm^{-1} is dominated by carbonyl vibrations of aromatic ring-opening products. The vibrations might be attributed to α,β-unsaturated γ-lactones, α,β-unsaturated acid peroxides or diaryl carbonates, and aryl or α,β-unsaturated esters. In combination with the vibration at 1790 cm^{-1}, the data reveal the presence of aryl or α,β-unsaturated carboxylic acid anhydrides. The presence of cis,cis-muconic acid as assumed by Bejan (2006) could not be confirmed.

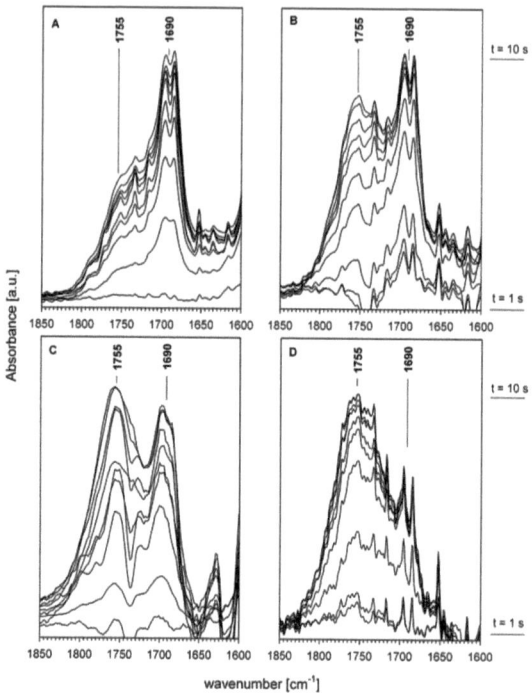

Figure 3.13: Evolution of the carbonyl stretching region of SOA from catechol at varying catechol/ozone ratios: A–1:0.5, B–1:1, C–1:1.5, D–1:5

The characteristics of the region of $\nu(C=O)$ stretching vibrations strongly depend on the precursor/ozone ratio (Figure 3.13). At low ozone concentrations (catechol/ozone = 1:0.5), the resulting carbonyl vibrations are dominated by aryl ketones and quinones or carboxylic acids at 1690 cm^{-1}. At concentrations of 1:1 and 1:1.5 (Figure 3.13 B and C), the formation of unsaturated non-aryl carbonyl vibrations is balanced with those of the aryl carbonyls. Finally, at high ozone concentrations (Figure 3.13 D), the region of $\nu(C=O)$ is dominated by unsaturated carbonyl stretching vibrations at 1755 cm^{-1}, resulting from aromatic-ring opening. Thus, the aromatic fraction of the resulting organic aerosol strongly depends on the ozone concentration as demonstrated by the characteristics of the $\nu(C=O)$ region.

The aerosol formation process from catechol as a precursor appears to start with the oxidation of the two hydroxyl groups, forming o-benzoquinone or derivatives of muconic acids. Ring-

3.3 SOA from catechol and guaiacol

Table 3.6: Assignment of infrared absorptions from SOA from catechol using the aerosol flow reactor

vibr. mode	[cm^{-1}]	assignment
$\nu(C=O)$	1790	carboxylic acid anhydrides, carboxylic acids, and esters from ring opening
$\nu(C=O)$	1755	aliphatic or α,β-unsaturated carbonyls from ring opening
$\nu(C=O)$	1690	aryl ketones, quinones, or olefinic carboxylic acids

opening processes, caused by oxidizing the carbonyls to carboxylic acids, and further oxidation results in highly oxidized compounds, in which the α and δ carbons are saturated (e.g. with hydroxyl groups). The entire aerosol spectrum is dominated by a broad absorption of hydroxyl groups, which also play a role in chemical bonding by their inter- and intramolecular hydrogen bonds. The presence of aromatic $\nu(C-H)$ and $\nu(C=C)$ indicates remaining aromatic structural elements inside the particle.

3.3.5 Spectroscopy of functional groups of SOA particles

While long-path absorption FTIR spectroscopy and the use of an aerosol flow reactor allows a deep insight into precursor transformation and SOA formation processes based on vibrational spectroscopy, these spectra always involve absorptions from the gas phase as well as the particulate phase. Therefore, ATR-FTIR spectroscopy and TPP-MS were applied to study functional groups and structural features of the particulate matter alone. SOA was prepared for FTIR and TPP-MS measurements at elevated concentrations, as described in the experimental section. This did not cause any identifiable discrepancies compared with reported natural samples or other HULIS models.

Different formation pathways of SOA from catechol lead to different chemical properties of the resulting aerosol (Figure 3.14). According to the long-path FTIR spectra, the major absorptions can still be assigned in the same manner. The aromatic structure of the SOA is indicated by the $-C=C-$ aromatic stretching vibration at 1620 cm^{-1}. A broad absorption in the range of the aromatic $C-H$ stretching (3100–3000 cm^{-1}) verifies this assumption. The aliphatic $C-H$ stretching vibration at 2960 cm^{-1} is less pronounced in the guaiacol aerosol. Also the absorptions at 860 and 740 cm^{-1}, originating from the $=C-H$ out-of-plane deformations, belong to the aromatic system of the organic aerosol.

SOA from catechol formed at 0 % relative humidity without irradiation by the solar simulator exhibits a C=O stretching vibration at 1716 cm^{-1}. This band indicates aryl aldehydes, α,β-unsaturated carboxylic acids, α,β-unsaturated aldehydes and α,β-unsaturated esters. Under simulated sunlight conditions, this band shifts to 1740 cm^{-1}, where the vibrations of saturated

3 SOA formation

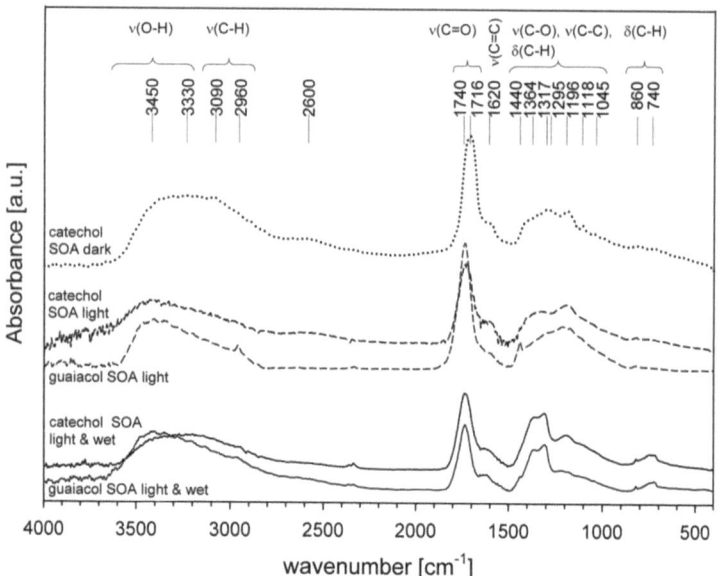

Figure 3.14: ATR-FTIR spectra of particulate matter of SOA from catechol and guaiacol at various ambient conditions

ketones, aldehydes, and esters are located. Aromatic or α, β-unsaturated esters in SOA formed in the dark might be indicated by the bands at 1716, 1295, 1196, and 1118 cm^{-1}. Carboxylic acids are indicated by the $\nu(C=O)$ at 1740 cm^{-1}, the $\nu(C-O)$ and $\delta(O-H)$ at 1364, 1317, and 1295 cm^{-1}, and the broad $\nu(O-H)$ from 3100 to 2500 cm^{-1}. The $\nu(O-H)$ absorption below 3000 cm^{-1}, coupled to the carbonyl stretch at low wavenumbers, observed for all samples, also gives hints at the formation of carboxylic acid dimers (Excoffon and Marechal, 1972; Florio et al., 2003).

Absorptions in the range of 3200–2500 cm^{-1} might belong to intramolecularly bonded orthophenols and the $O-H$ stretching vibration of carboxylic acids. The broad structured absorption between 1400 and 1000 cm^{-1} implies $O-H$ deformation and $C-O$ stretching vibration combinations of aliphatic and aromatic alcohols. The band at 1196 cm^{-1} observed for all SOA samples might, apart from structural features, also belong to the $C-O$ stretching vibration of aromatic ethers or phenols. Hence, the absorption at 1045 cm^{-1}, which is masked by ozone in the gas-phase long-path absorption spectra, might belong to the aliphatic $C-O$ stretching vibration of aliphatic-aromatic ethers. Similar absorptions of carbonyls and aromatic rings were found for photo-degraded tannic acid as a model for HULIS (Cowen and Al-Abadleh, 2009) or for SRFA (Hatch et al., 2009).

3.3 SOA from catechol and guaiacol

ATR spectra of SOA from guaiacol show the asymmetric stretch (2960 cm^{-1}) and asymmetric deformation (1440 cm^{-1}) of the $-CH_3$ group. The intensity of those vibrations is reduced under wet conditions, because the $-CH_3$ group is destroyed by reaction with the OH radicals, a process that is commonly known from the atmospheric degradation of aliphatic compounds (Seinfeld and Pandis, 2006). All other bands can be assigned in the same way as for the catechol experiment.

By interpreting the ATR-FTIR spectra of the different SOA formed, the infrared group frequencies could be assigned to the functional groups listed in table 3.7.

3 SOA formation

Table 3.7: Assignment of infrared absorptions of SOA from catechol and guaiacol using ATR-FTIR

vibr. mode	[cm^{-1}]	assignment	catechol SOA[1]			guaiacol SOA[1]	
			dark	light	light & wet	light	light & wet
$\nu(O-H)$	3450	intramolecular	m	m	m	m	m
$\nu(O-H)$	3330	various	m	m	m	m	m
$\nu(C-H)$	3090	aromatic =CH	m	w	w	w	w
$\nu(C-H)$	2960	asym. aliphatic $-CH$ and $-CH_x$	m	w	w	m	w
$\nu(C-H)$	3200 - 2500	intramolec. phenols or carboxylic acids	m	m	w	w	w
$\nu(=O\cdots H-)$	2600	R/Ar-COOH dimers [a]	m	w	w	w	w
$\nu(C=O)$	1740	olefinic and saturated		s	s	s	s
$\nu(C=O)$	1716	aryl and unsaturated	s				
$\nu(C=C)$	1620	aromatic, olefinic	m	m	m	m	m
$\delta(C-H)$	1440	aliphatic $-CH$ and $-CH_x$	m	w	w	w	w
$\nu(C-O)$	1364	carboxylic acids or alcohols	m	m	m	s	s
$\nu(C-O), \delta(O-H)$	1317	carboxylic acids or esters	m	m	m	s	s
$\nu(C-O), \delta(O-H)$	1295	carboxylic acids or esters	m	m	m		
$\nu(C-O)$	1196	Ar-OH, Ar-O-Ar, R-COO-R	m	m	m	m	m
$\nu(C-O)$	1118	of R-O-Ar [a] or esters	m	w	w	w	w
$\nu(C-O)$	1045	of R-O-Ar or R-O-R [a]	m	w	w	w	w
$\delta(C-H)$	860	=C-H out-of-plane def.	w	w	w	w	w
$\delta(C-H)$	740	=C-H out-of-plane def.	w	w	m	w	m

[a] Ar - representing conjugated or aromatic structural elements; R - representing aliphatic or substituted olefinic structural elements
[1] w = weak, m = medium, s = strong

3.3 SOA from catechol and guaiacol

TPP-MS (for details see 2.6) was used to determine oxygen-containing functional groups. The background-corrected TPP mass spectra of the five aerosol types are well structured for the masses 17 (OH), 28 (CO) and 44 (CO_2) (Figure 3.15). Peaks below 150 °C were not assigned because of outgassing of physically adsorbed molecules, like H_2O, N_2, CO and CO_2, which might pollute the signals of pyrolyzing functional groups. Based on the thermal stability of oxygen-containing functional groups and their fragments (Muckenhuber and Grothe, 2006), peak maxima and relative peak intensities have been assigned to the different aerosol types (Table 3.8).

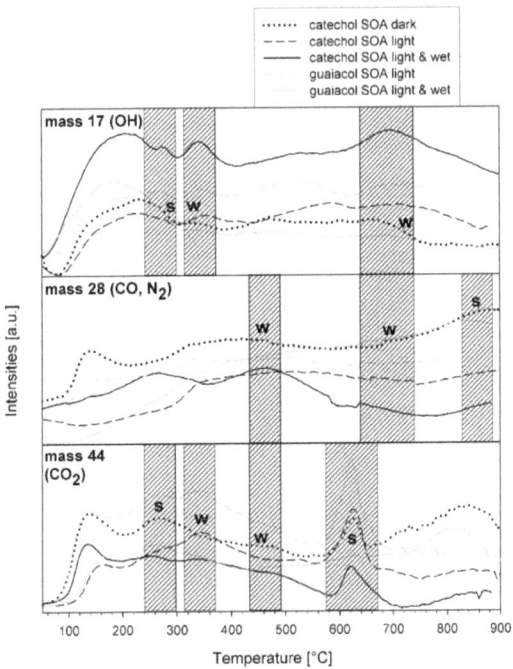

Figure 3.15: TPP-MS signals of the main masses for oxygen-containing functional groups observed as a function of temperature for the five different SOA

Two thermal instabilities of carboxylic acids occur at 270 and 340 °C. There are small differences in the relative intensities of those two acids. The m/z = 44 signal assigned to lactones or esters is very intense for all aerosol particles except for SOA from guaiacol formed under light and wet conditions. Further carboxylic anhydrides, phenols, carbonyls, and quinones are visible in the TPP mass spectra. Ethers might decompose at 680 °C, but the respective m/z = 28 signal is rather small.

65

3 SOA formation

Table 3.8: Strength of the TPP-MS signals of functional groups at their respective decomposition temperatures for the five different SOA

Functional groups	carboxylic acids	carboxylic anhydrides	lactones or esters	phenols	ethers	carbonyls and quinones
Decomposition temperatures	270	320	460	620	680	680	860
SOA catechol dark[1]	s	w	w	s	w	w	s
SOA catechol light[2]	↓	=	↑	=	=	=	↓
SOA catechol light and wet[2]	↓	↑	=	=	↑	=	↓
SOA guaiacol light[2]	=	↑	↑	=	=	=	–
SOA guaiacol light and wet[2]	↓	↑	=	↓	=	=	↓

[1] Intensity of the mass signals: s = strong, w = weak
[2] The relative intensities are given in relation to SOA from catechol formed in the dark, and their intensities are indicated by the following symbols: stronger (↑), equal(=), weaker (↓), not observed (−).

The thermal analysis of functional groups is in good agreement with the vibrational spectra, confirming that highly oxidized functional groups play a major role, but demonstrating that lower oxidized groups are also present in the aerosol particles. When the oxidation reaction is enhanced by changed environmental conditions, carboxylic acids, which decompose at higher temperature (320 °C), increase. However, carboxylic acids decomposing at lower temperatures (270 °C) as well as carbonyls and quinones decrease. Minor increases were observed for carboxylic anhydrides and phenols. The amount of lactones or esters appears to be unchanged for different simulated environmental conditions.

3.3.6 Optical properties

The diffuse-reflectance UV/VIS spectra of the different aerosol particles from catechol and guaiacol are dominated by an extremely broad absorption up to 600 nm (Figure 3.16(c)), which is in good agreement with the brown color of these samples (Figure 3.16(a) and 3.16(b)). Individual electronic transitions can barely be distinguished due to this broad absorption, underlining the fact that numerous types of conjugated bonds exist in the SOA. The main absorptions of the two precursors catechol and guaiacol in the UV range are located at about 220 and 275 nm. They are related to the $\pi \rightarrow \pi^*$ transition of the aromatic system and the $n \rightarrow \pi^*$ transition of the lone pairs of the hydroxyl group. Within the above-mentioned broad absorption of the

3.3 SOA from catechol and guaiacol

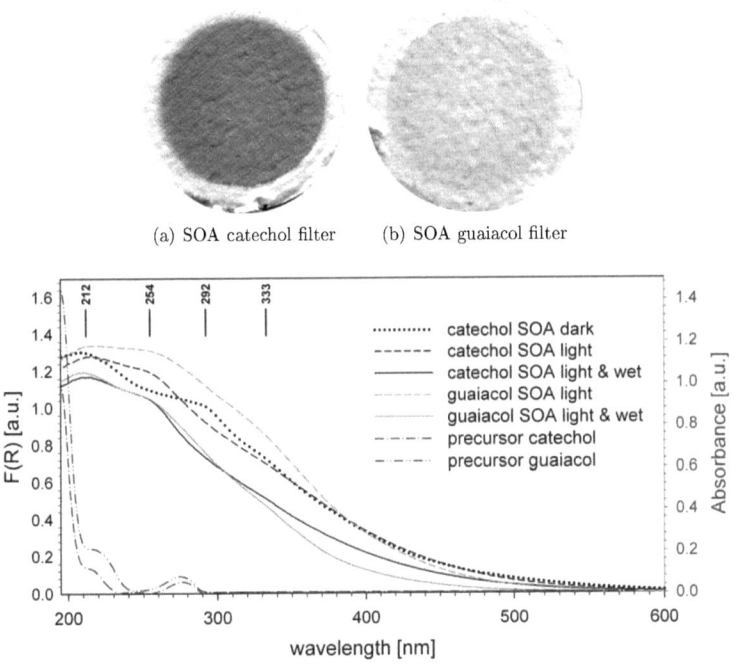

(a) SOA catechol filter   (b) SOA guaiacol filter

(c) Diffuse-reflectance UV/VIS absorbance spectra of the precursors

Figure 3.16: Diffuse-reflectance UV/VIS spectra of SOA from catechol and guaiacol, absorbance spectra of the precursors, and images of SOA samples

organic aerosol particles, three main absorptions occur at 212, 254, and 333 nm, which are present in all different types of organic aerosol from catechol and guaiacol. One additional absorption at 292 nm is only present in SOA from catechol formed in the dark. This transition seems to be destroyed by UV/VIS radiation.

The UV/VIS spectra are in good agreement with other aromatic-precursor oxidation studies (Gelencser et al., 2003; Hoffer et al., 2004). Especially the $\pi \rightarrow \pi^*$ electron transition at about 260 nm is similar to that of natural HULIS samples (Baduel et al., 2009).

The samples shown in figure 3.16(a) and 3.16(b) were prepared at 25 % relative humidity using simulated sunlight. As visible in figure 3.16(c), the catechol SOA is a stronger absorber towards higher wavelengths than the guaiacol SOA.

All traces shown in figure 3.16(c) exhibit a nearly linear decrease towards the red end of the spectrum if the absorbance is plotted on a logarithmic scale.

3.3.7 Ultra-high-resolution mass spectra of SOA from catechol and guaiacol

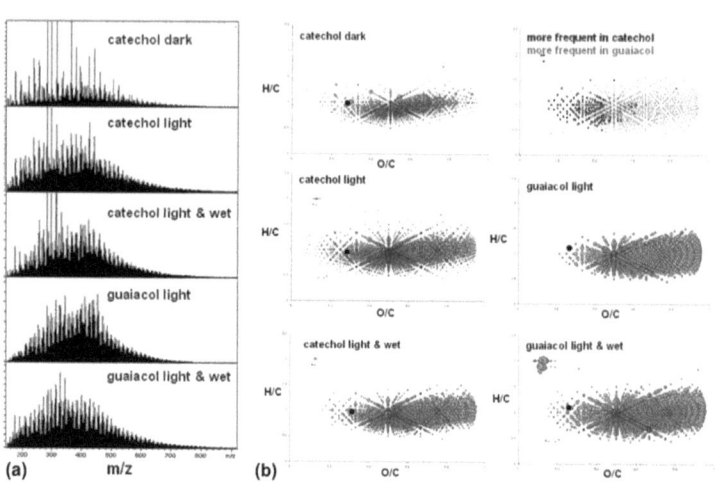

Figure 3.17: Ultra-high-resolution mass spectra (a) and Van Krevelen diagrams (b) of the different SOA samples

The use of ultra-high-resolved mass spectroscopy allowed to determine single compounds of the organic aerosols and to classify the SOA according to Van Krevelen diagrams (H/C ratio versus O/C ratio).

The methanol extracts of the filter samples exhibit a Gaussian distribution of the peaks between 150 and 900 m/z (Figure 3.17 (a)) in negative-mode electrospray ICR-FT/MS. The main region is between 200 and 450 m/z. The polymers show thousands of signals for which individual CHO elementary compositions were computed. Due to the high resolution of the mass signals, the composition can be calculated based on the mass defect of the individual elements. The chemical diversity of the samples formed in the presence of light is significantly increased for both catechol- and guaiacol-based mixtures; humidity, however, did not have the same impact. The main m/z ratio of 200–450 of organic molecules in the SOA particles is closer to natural HULIS samples than SRFA (Graber and Rudich, 2006), and is comparable to other SOA samples from photo-oxidation (Kalberer et al., 2004). The mass spectra in figure 3.17 (a) also exhibit a polymeric structure, revealed by the regular appearance of strong signals. This is also in good agreement with the reported polymeric structures of SOA (Kalberer et al., 2004).

3.3 SOA from catechol and guaiacol

The distribution of the peaks in ICR-FT/MS is very similar for the catechol- and guaiacol-aerosol extracts. A detailed description on the elementary composition level shows signals that are typical of the catechol or guaiacol precursors.

Hundreds of calculated elemental compositions were transformed into atomic H/C and O/C ratios for the plotting of Van Krevelen diagrams (Hertkorn et al., 2007, 2008). Based on only one molecular precursor, the resulting aerosols cover a major part of the allowed CHO compositional space. Previous ICR-FT/MS analyses on organic aerosols obtained from chamber experiments based on α-pinene ozonolysis permitted the differentiation of various monomers to oligomers (Reinhardt et al., 2007); in that case, however, the mixtures were characterized with H/C values ranging continuously from 0.5 to 1.5 and O/C values ranging from 0.3 to 1 with gradual changes in peak intensity with increasing oxygen content. This behavior is typical of the catalyzed oxidative polymerization of these phenols to polyphenols, as described in the early days of soil humic-substance chemistry (Stevenson, 1994). Organic aerosol from catechol and guaiacol is typically characterized by high oxygen-to-carbon ratios (Figure 3.17 (b)), indicating the presence of highly-oxidized benzenes or conjugated olefins. Catechol-based mixtures also differ significantly from guaiacol (Figure 3.17 (b)), with signals of higher intensity being affiliated a lower oxygen content and higher aromaticity. The O/C ratio located between 0.3 and 1 is in good agreement with the reported O/C ratios for LV-OOA (low-volatile oxidized organic aerosol) and SV-OOA (semi-volatile OOA) (Jimenez et al., 2009). Furthermore, the mean value of 0.6–0.7 fits the described oxidation state of atmospheric HULIS very well, although the measured H/C ratio of about 1 is lower than the reported value of 1.6–1.7 (Graber and Rudich, 2006). The H/C ratio is in good agreement with the commonly used proxy SFRA (Dinar et al., 2006).

Based on the chemical structure of the precursors, hydrogen-to-carbon ratios above 1 and high oxygen contents can only be explained by condensation reactions and the addition of hydroxyl groups to the unsaturated carbon structure, which is confirmed by the broad $\nu(O-H)$ absorption in the FTIR spectra.

The use of ultra-high-resolution mass spectroscopy (example of data is given in figure 3.18) gave access to the calculation not only of O/C and H/C ratios, but also the carbon oxidation state $\overline{OS_C}$ according to equation 1.1. The calculated ranges of these parameters are summarized in table 3.9.

3 SOA formation

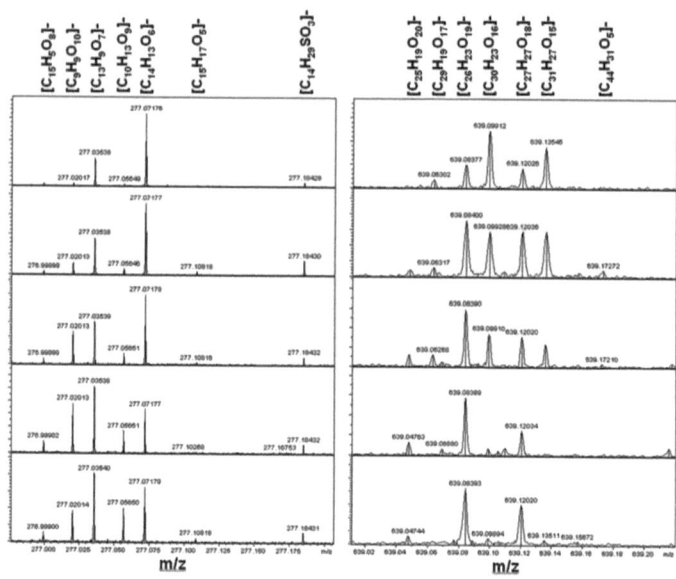

Figure 3.18: Examples of elemental compositions determined by ICR-FT/MS

Table 3.9: Calculated ranges of mass-spectroscopic values of SOA

aerosol	O/C ratio	H/C ratio	$\overline{OS_C}$	n_C
SOA from catechol	0.44 - 1	0.57 - 1.04	-0.13 - 0.91	11 - 31
SOA from guaiacol	0.44 - 1.2	0.57 - 1.09	-0.13 - 1.6	10 - 27

4 Molecular processing

Model halogenation was used to determine the influence of some common halogen species on the different types of organic aerosol without involving the complex reaction scheme of halogen activation and release from heterogeneous surfaces.

To study the aerosol-halogen interaction process, the well-characterized organic aerosols (see chapter 3) were exposed to gaseous molecular halogens, which were photolyzed by UV/VIS irradiation. These model halogenations were compared to SOA halogenation processes by halogens released from sea-salt aerosol and from a simulated salt pan (see chapter 5).

4.1 Experimental setup

Model halogenations were performed using the 700 L aerosol smog chamber (see chapter 2). For the preparation of bromine and chlorine radicals, molecular bromine (Fluka, 196050, puriss., > 99.0 %) and molecular chlorine (Aldrich, 295132-454G, > 99.5 %) were used.

All organic aerosols were formed by gas-to-particle partitioning in situ in the aerosol smog chamber under simulated sunlight, ozone, and two different relative humidities (0 and 25 %) as described in chapter 3.

After one hour of aerosol formation, a stoichiometric amount (1 precursor molecule to 1 halogen molecule) of molecular bromine or chlorine was added into the aerosol smog chamber.

Experiments to study the influence on the aerosol size distribution were performed using 300 ppb aerosol precursor, 1 ppm ozone, and 300 ppb molecular halogen. To achieve an adequate amount of particulate matter for the physicochemical analysis, 5 ppm of the aerosol precursor, 20 ppm of ozone, and, after one hour of aerosol formation, 5 ppm of molecular halogen were added.

The dosing of the halogens is described in detail in section 2.9.

4.2 Influences on aerosol size and particle number distributions

The aerosol size distribution of the organic model aerosols is strongly influenced by the interaction with the halogen (Figure 4.1). The addition of halogen causes an increase of the mean particle diameter, which was observed for nearly all aerosols studied. The observed increase is stronger for the reaction with chlorine than for the reaction with bromine. While this result fully describes the behaviour of catechol-derived SOA, SOA from guaiacol and α-pinene exhibit some special features. In general, influences of chlorine on the aerosol size distributions are stronger than of bromine. The reaction with SOA from α-pinene also is an exception in this case.

The reaction of chlorine with SOA from guaiacol leads to a bimodal size distribution with a second maximum at small particle diameters. Due to the fact that a methyl-ether group is the only difference between the catechol and the guaiacol precursor, the formation of this maximum is related to the methyl-ether group. This second mode is not observed for the reaction with bromine. The formation of this mode appears to be caused by a very high supersaturation of the remaining gas-phase species of the oxidized organic precursor. In general, the formation of low-volatile compounds from gas-phase species is indicated by the increase of the main particle diameter. In case of guaiacol, a large amount of oxidized precursor appears to be available during the reaction with chlorine. The reaction of these precursors with chlorine species causes a supersaturation too high to be reduced only by condensing on already existing aerosol particles. Therefore, condensation of the available low-volatile material leads to a second maximum of the size distribution.

Although the mean particle diameter increases by the reaction of halogens with the organic aerosol from catechol and guaiacol, another phenomenon occurs during the reaction of α-pinene SOA with the halogen species. The mean diameter increases immediately after addition of the halogen, but remains constant (at 0 % relative humidity) or decreases slightly (at 25 % relative humidity) afterwards. The reduction of the mean particle diameter is stronger for the reaction with bromine than for the reaction with chlorine. The shift of the aerosol size distribution maximum to smaller sizes for SOA from α-pinene after reaction with halogen species might be caused by the abstraction of hydrogens from the aliphatic $-CH_3$ and $-CH_2-$ groups.

Thus, the evolution of the aerosol size distribution for different organic aerosols reacting with halogens is not only depending on the halogen (chlorine or bromine) and the relative humidity, but also on the chemical composition of the organic aerosol. The chemical structure of the organic precursor, and therefore the structural and functional features of the resulting secondary organic aerosols, strongly influence the evolution of the mean particle diameter. The

4.3 Changes in the vibrational gas-phase and particle-phase spectra

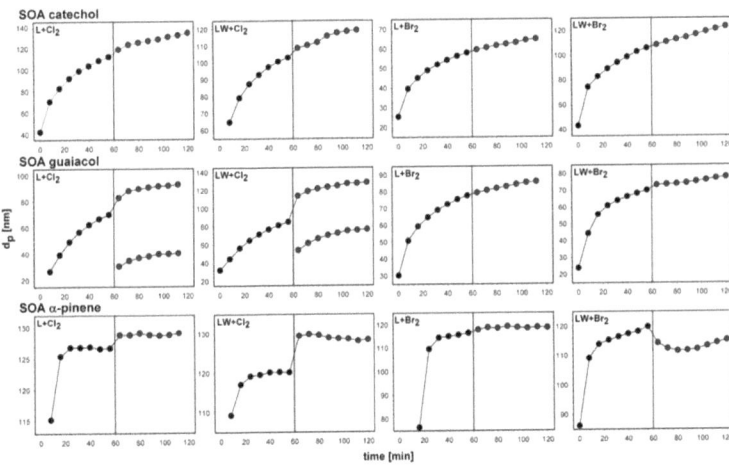

Figure 4.1: Change of the mean particle diameter of organic aerosols by the reaction with gaseous halogens (L: SOA formed under simulated sunlight at 0 % relative humidity; LW: SOA formed under simulated sunlight at 25 % relative humidity), with the vertical lines marking the injection of the halogen after 60 minutes aerosol formation.

influence of the change of the hygroscopicity of the organic-aerosol surface or the available oxidized precursor on the change of the aerosol size distribution is not yet understood.

The increase of the mean aerosol size, observed for nearly all samples, might be caused by the formation of further low-volatile compounds due to the halogenation of gaseous organic species and condensation on the existing aerosol particles. The formation of the second particle mode for SOA from guaiacol can be related to the methyl-ether group, either to an abstraction of the group or to supersaturation by the fast generation of oxidized compounds. An additional abstraction process causes the decrease of the mean particle diameter, observed for SOA from α-pinene. Hydrogen abstraction and degradation of aliphatic structural elements could be the reason for the different behavior of SOA from α-pinene.

4.3 Changes in the vibrational gas-phase and particle-phase spectra

During the reaction of the two halogens with the organic aerosols several changes of gaseous species occur, identified by long-path absorption spectroscopy (Figure 4.2). Due to the rather low spectral resolution compared to commonly used resolutions for gas-phase infrared spectra,

4 Molecular processing

only a few gas-phase species were identified by their vibrational-rotational spectrum. For species which could not be resolved, the characteristic P(Q)R-structure of their absorptions was used to ascertain their gaseous state.

After adding the halogens, a strong increase in CO_2 is monitored (2349 and 668 cm^{-1}). At the same time, ozone, which was not totally consumed by the aerosol formation process, decreases (1044 cm^{-1}). Furthermore, HCl or HBr, respectively, is formed. Gaseous HCl, with $\nu(H-Cl)$ at 2887 cm^{-1}, is formed to a greater extent than HBr, with $\nu(H-Br)$ at 2559 cm^{-1}. While CO_2 and HCl or HBr increase, aromatic and aliphatic $\nu(C-H)$ decrease.

This is observed for all three model aerosols: SOA from catechol (Figure 4.2(a)) only exhibits the aromatic $\nu(C-H)$ at 3050 cm^{-1}, SOA from guaiacol (Figure 4.2(b)) the aromatic $\nu(C-H)$ at 3050 cm^{-1} as well as the aliphatic $\nu(C-H)$ of the $-CH_3$ of the methyl-ether group between 2800 and 3000 cm^{-1}. The aliphatic $\nu(C-H)$ of the $-CH_3$ groups of SOA from α-pinene (Figure 4.2(c)) also decreases under the influence of halogens. The prominent absorption at 740 cm^{-1} of SOA from guaiacol ($\delta(C-H)_{out-of-plane}$ (Ofner et al., 2011)) also seems to decrease. Furthermore, CO with its absorption at 2143 cm^{-1} is formed for all different aerosols and under all ambient conditions.

In contrast to all other experiments, CO was consumed by further reactions for the reaction of SOA from catechol with chlorine at 0 % and 25 % relative humidity. Furthermore, an additional absorption at 850 cm^{-1} occurs. This absorption shows the P-and-R branch structure of a gaseous species and can be assigned to phosgene (point group C_{2v}). The associated $\nu(C=O)$ at 1827 cm^{-1} could not be identified because the spectral region of the carbonyl stretching vibrations was not recorded.

At 1106 cm^{-1}, an absorption of a gas-phase species with PQR-structured branches is visible in each long-path absorption spectrum. This absorption sometimes decreases or increases. The assignment of this vibration to a specific molecule was not possible. It might be formic acid, which as a very strong $\nu(C-O)$ at 1105 cm^{-1}. The strong τ vibration at 638 cm^{-1} might be concealed by CO_2, and all other vibrations are weak or within the carbonyl-stretching region (Shimanouchi, 1972). All other absorptions were assigned to gaseous species based on reported group frequencies (Siebert, 1966). A list of the most significant changes observed in the long-path absorption infrared spectra during the reaction of the model SOA with halogens is given in table 4.1.

4.3 Changes in the vibrational gas-phase and particle-phase spectra

Table 4.1: Assignment of decrease and increase of infrared absorptions of the three different SOA reacting with halogens measured by long-path absorption infrared spectroscopy

vibr. mode	[cm^{-1}]	assignment	SOA catechol[1]		SOA guaiacol[1]		SOA α-pinene[1]	
			L[2]	LW[2]	L	LW	L	LW
Reaction with chlorine								
$\nu(C-H)$	3050	aromatic $C-H$	wD	wD				
$\nu(C-H)$	3000–2880	aliphatic $C-H$			sD	sD	sD	sD
$\nu(H-Cl)$	2887	gaseous HCl	sF	wF	wF	wF	wF	wF
$\nu(C=O)$	2349	Fermi resonance of CO_2	sF	sF	sF	sF	sF	sF
$\nu(C=O)$	2143	CO	wD	wD	wF	wF	wF	wF
$\nu(C-O)$	1106	possibly formic acid	sD	sD	sD	sD	wF	wF
$\nu_{as}(C-Cl)$	851	possibly phosgene	wF	wF				
$\delta(O=C=O)$	668	CO_2	sF	sF	sF	sF	sF	sF
Reaction with bromine								
$\nu(C-H)$	3050	aromatic $C-H$	sD	wD				
$\nu(C-H)$	3000–2880	aliphatic $C-H$			wD	wD	sD	wD
$\nu(H-Br)$	2559	gaseous HBr	wF	wF	sF	wF	wF	wF
$\nu(C=O)$	2349	Fermi resonance of CO_2	sF	sF	sF	sF	sF	sF
$\nu(C=O)$	2143	CO	sF	wF	sF	sF	sF	wF
$\nu(C-O)$	1106	possibly formic acid	wF		wD	wF	sF	wF
$\delta(O=C=O)$	668	CO_2	sF	sF	sF	sF	sF	sF

[1] wF = weak formation, sF = strong formation, wD = weak destruction, sD = strong destruction
[2] L = SOA formed at 0 % relative humidity and simulated sunlight; LW = SOA formed at 25 % relative humidity and simulated sunlight

4 Molecular processing

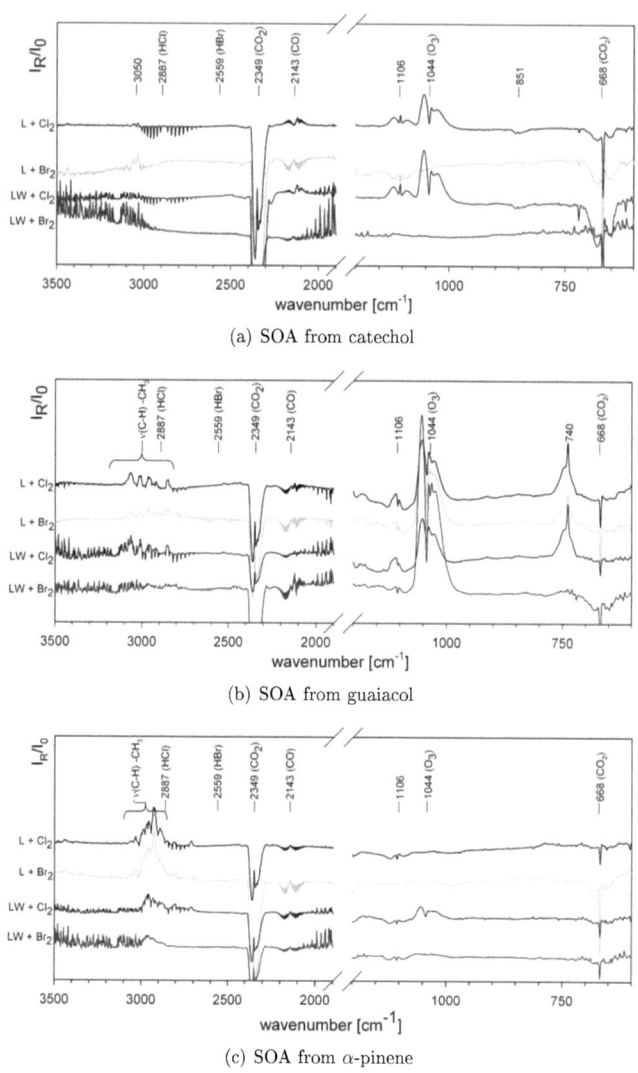

Figure 4.2: Long-path infrared spectroscopy of the heterogeneous reaction of the different organic aerosols with molecular halogens; L: SOA formed under simulated sunlight at 0 % relative humidity; LW: SOA formed under simulated sunlight at 25 % relative humidity)

4.4 Changes in the vibrational spectra of the particulate phase by reaction with halogens

To investigate the transformation of functional groups, ATR-FTIR spectroscopy of the particulate phase of the aerosol deposited on KRS5 crystals was performed (Figure 4.3). The infrared spectra of the particulate phase of the model aerosols have been reported in section 3.2.4 (SOA from α-pinene) and section 3.3.5 (SOA from catechol and guaiacol).

After reacting with halogens, significant changes occur in the infrared spectra of the aerosols. While the $\nu(C=O)$ absorption is the most prominent for the untreated aerosols, this vibration is less intensive after the reaction. The spectra of all SOA exhibit two bands at 1372 and 1303 cm^{-1} after treatment with halogens. In the case of catechol-derived SOA these bands are the most prominent in the spectrum. According to the literature (Shevchenko, 1963; Baes and Bloom, 1989; Ofner et al., 2011), these absorptions can be assigned to carboxylic acids, carboxylic acid salts, and other $\nu(C-O)$ vibrations of aromatic or aliphatic $R-OH$, $R-COO-R$, or $R-O-R$. Flett (1962) reports a strong absorbance for the symmetric stretching vibration of $-CO_2^-$ of carboxylic acid salts of small molecules close to this region. The lowered intensity of the carbonyl stretch vibration, as observed in the present study, was reported by Rontu and Vaida (2008). In their investigation of perfluoro-carboxylic acids, the most prominent absorption between 1000 and 1400 cm^{-1} seems to be the $\nu(C-F)$, but also the $\nu(C-O)$ is more intensive than the $\nu(C=O)$. Hence, halogens seem to significantly influence carboxylic acids of SOA by degradation or by changing their vibrational properties as α- or β-substituent.

Absorptions between 700 and 800 cm^{-1} have been reported for SOA from catechol as well as from guaiacol (section 3.3.5 or Ofner et al. (2011)), if the SOA was formed at 25 % relative humidity and simulated sunlight. For α-pinene, the lowest observed absorption is located at 854 cm^{-1} (section 3.2.4 or Sax et al. (2005)). The ATR spectra of treated SOA from catechol also exhibit strong absorptions between 800 and 700 cm^{-1}, also for SOA formation and reaction at 0 % relative humidity. This absorption can be assigned to alcohols or carboxylic acids, but in the case of treatment with chlorine also to the $\nu(C-Cl)$ vibration (Socrates, 1980). Similar, but weaker, absorptions were observed for SOA from guaiacol and α-pinene. During the reaction with bromine, an additional absorption was observed at 609–602 cm^{-1} for all three kinds of SOA, which can be assigned to the $\nu(C-Br)$ vibration. Therefore, the most significant changes of the vibrational spectra of organic aerosols through treatment with halogens are observed for the reaction with bromine, which influences carboxylic acids, and results in absorptions in the $\nu(C-Br)$ region.

4 Molecular processing

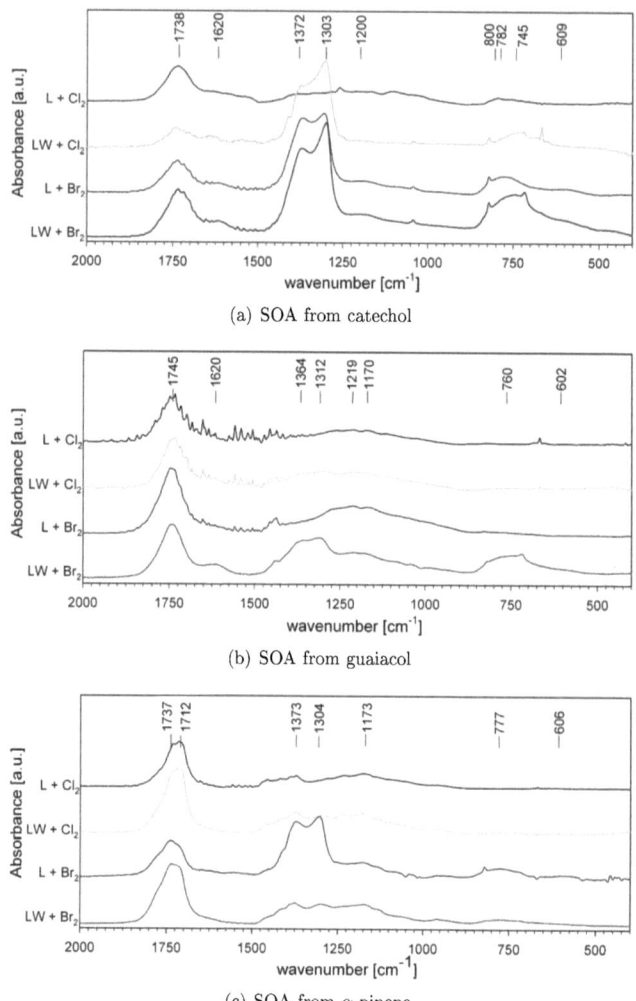

Figure 4.3: ATR-FTIR spectra of the particulate phase of the aerosol after reaction with halogens; L: SOA formed at 0 % relative humidity and simulated sunlight; LW: SOA formed at 25 % relative humidity and simulated sunlight)

Table 4.2: Assignment of absorptions observed in the ATR spectra of the three different SOA reacting with halogens

vibr. mode	[cm^{-1}]	assignment	SOA catechol[1]		SOA guaiacol[1]		SOA α-pinene[1]	
			L[2]	LW[2]	L	LW	L	LW
Reaction with chlorine								
$\nu(C-O)$	1380-1300	carboxylic acids or their salts		s			m	
$\nu(C-Cl)$	800-700		w	m	w	w		w
Reaction with bromine								
$\nu(C-O)$	1380-1300	carboxylic acids or their salts	s	s	w	m	s	m
$\nu(C-Br)$	609-602		m	s	w	s	m	w

[1] w = weak, m = medium, s = strong
[2] L = SOA formed at 0 % relative humidity and simulated sunlight; LW = SOA formed at 25 % relative humidity and simulated sunlight

4.5 Identification of halogenated compounds in the particle phase

TPP-MS was used to determine the presence of halogenated species in the particle phase (Figure 4.4). For each halogen atom (chlorine and bromine), the masses of the two stable isotopes were used to calculate the isotopic ratio. Halogenated compounds could be measured in thermal regions were decomposing compounds exhibit an isotopic ratio close to the natural one, which is 0.32 for chlorine ($^{37}Cl/^{35}Cl$) and 0.97 for bromine ($^{81}Br/^{79}Br$). The mass ratios were plotted as a function of temperature to determine the release of atomic chlorine and bromine from the SOA samples.

Most of the samples exhibit strong peaks of these masses between 100 and 400 °C. These peaks correlate well with the isotopic ratio of bromine. The correlation with chlorine is weak, but a significant decrease of the mass ratio close to the isotopic ratio could also be determined. The poor correlation between the $^{37}Cl/^{35}Cl$ ratio and the isotopic ratio can be explained by a lower amount of bound chlorine and by a higher influence of other fragments, which thermally decompose, in this mass range. All halogens exhibit a rather low thermal stability, and therefore a weak bond with the macromolecular structure of the particulate matter.

Hydrohalogens seem to be the main gases released from the samples, as indicated by figure 4.5. The release is caused by either the desorption of absorbed hydrohalogens or by the de-

4 Molecular processing

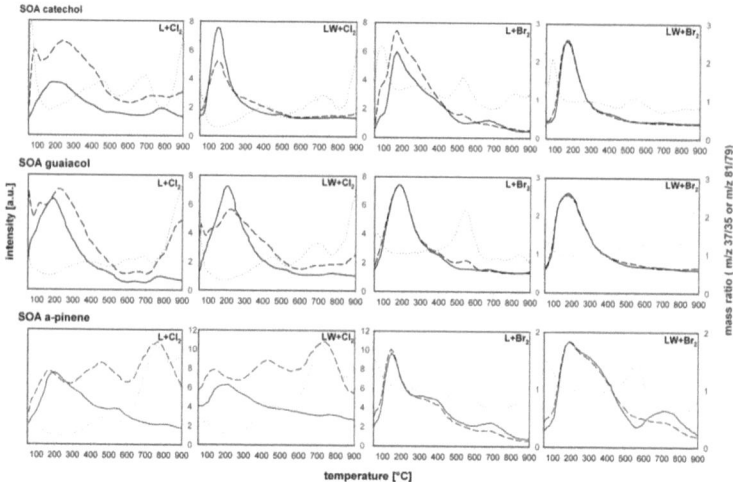

Figure 4.4: TPP-MS spectra of the masses 35, 37, 79 and 81, representing ^{35}Cl, ^{37}Cl, ^{79}Br, and ^{81}Br, and the calculated mass ratio (dotted) to determine thermal regions were the isotopic ratios can be found (L: SOA formed at 0 % relative humidity and simulated sunlight; LW: SOA formed at 25 % relative humidity and simulated sunlight)

composition of halogen-containing functional groups and subsequent recombination of halogen atoms with hydrogen atoms. Other species, like phosgene, were not be observed in the mass spectra. While Gaussian desorption peaks would indicate a purely physical desorption of the adsorbed species, more structured peaks with no Gaussian shape point to an influence of different thermally stable functional groups. Therefore, most of the desorption processes of halogens shown in figure 4.4 can be assigned to fragments of covalently-bonded halogens.

The reactions with bromine result in a second peak at higher thermal stabilities (350 °C). These two peaks are most prominent for SOA from α-pinene. Thus, two different kinds of bonding of bromine to the macromolecular structure of the SOA were identified in the case of SOA containing aliphatic structural elements.

The release of the halogen species, verified by the isotopic ratios, in the thermal region where carboxylic acids desorb (Muckenhuber and Grothe, 2006) could explain the appearance of the prominent absorptions in the ATR spectra, which are linked to carboxylic acids or carboxylic acid salts. Hence, an interaction between halogens and the carboxylic acids of SOA is likely.

The kind of halogenation was also studied using ultra-high-resolution mass spectroscopy of filter samples after reaction with the halogens at 0 % relative humidity and simulated sunlight (Figures 4.6, 4.7 and 4.8). Halogenation products occurred at nearly all m/z and O/C ratios.

4.5 Identification of halogenated compounds in the particle phase

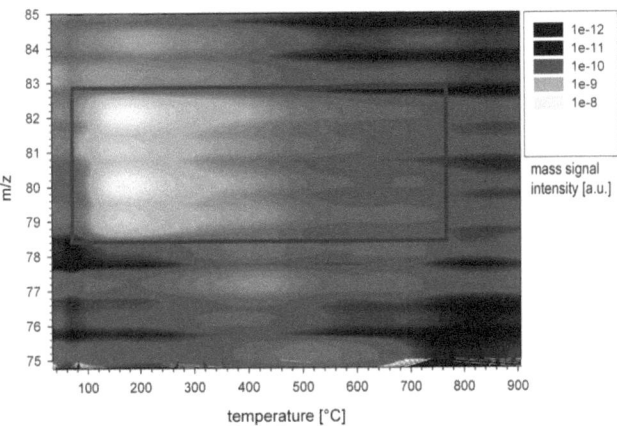

Figure 4.5: TPP mass spectra of SOA from catechol (LW) between 75 and 85 amu: The masses of the halogen atoms (^{79}Br and ^{81}Br) correlate with those of HBr (^{80}M and ^{82}M).

Table 4.3: Degree of halogenation of SOA formed at 0 % relative humidity and simulated sunlight

Elemental composition of molecules	SOA catechol	SOA guaiacol	SOA α-pinene
CHO	100 %	100 %	100 %
CHOCl	3 %	15 %	33 %
CHOBr	30 %	53 %	66 %

The reaction of bromine with SOA is more efficient than the reaction with chlorine (Table 4.3). The reactions of chlorine or bromine with catechol or guaiacol lead to compounds with higher H/C ratios compared with the original precursor (H/C up to 2). While the chlorinated compounds are located at rather low m/z ratios (250–450), brominated compounds with an H/C ratio of about 2 could be found from 450 to 850 m/z. Furthermore, an increase in halogenation could be observed from purely aromatic structures to aliphatic structures (Table 4.3). While this increase seems to be in contrast to the halogen-containing vibrations observed by ATR-FTIR spectroscopy, the observed bands are not only correlated to the amount of halogenated compounds, but they also depend on the chemical neighbourhood of the bonded halogens.

4 Molecular processing

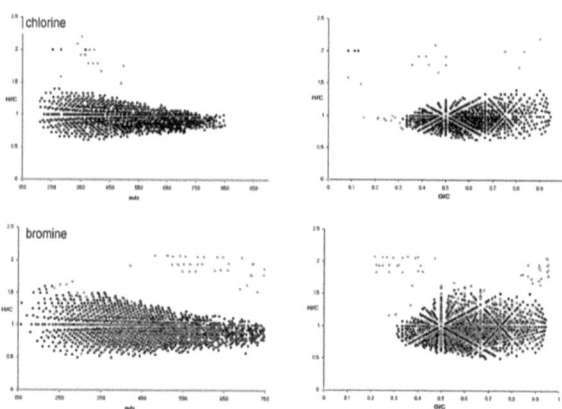

Figure 4.6: Van Krevelen diagrams of halogenated SOA from catechol formed at 0 % relative humidity and simulated sunlight

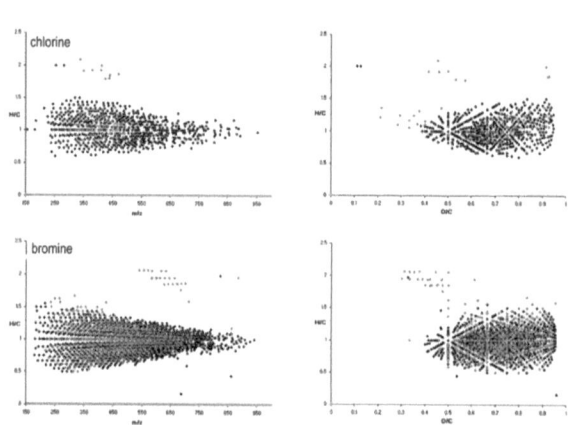

Figure 4.7: Van Krevelen diagrams of halogenated SOA from guaiacol formed at 0 % relative humidity and simulated sunlight

4.5 Identification of halogenated compounds in the particle phase

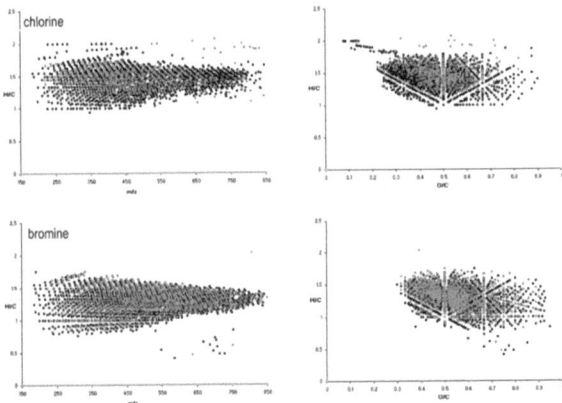

Figure 4.8: Van Krevelen diagrams of halogenated SOA from α-pinene formed at 0 % relative humidity and simulated sunlight

4.6 Parameters of processed SOA calculated from ICR-FT/MS spectra

Based on the available data acquired by ultra-high-resolution mass spectroscopy (example given in figure 4.9(b)), the main parameters of the processed SOA were calculated. The observed mass ranges of the processed SOA (example given in figure 4.9(a)) are comparable to the unprocessed SOA (see section 3.3.7).

The main ranges of these parameters are summarized in table 4.4. While the available data only gave access to carbon numbers of up to 16, higher numbers of carbon atoms is also likely (see table 3.9). This assumption is based on the comparison of figure 3.17 (a) with figure 4.9(a).

Table 4.4: Main parameters of processed SOA, calculated from ICR-FT/MS spectra

aerosol	O/C ratio	H/C ratio	$\overline{OS_C}$	n_C
SOA catechol + Cl	0.31–0.67	0.81–1.08	-0.19–0.25	12–16
SOA catechol + Br	0.55–1.2	0.9–1.67	0–1.6	5–11
SOA guaiacol + Cl	0.31–0.67	0.6–1.08	-0.13–(-0.33)	12–16
SOA guaiacol + Br	0.6–1.2	0.9–1.67	0–1.6	5–10
SOA α-pinene + Cl	0.33–0.54	1.31–1.67	-0.93–(-0.15)	13–15
SOA α-pinene + Br	0.42–0.83	1.18–2	-0.5–0.6	6–12

4.6 Parameters of processed SOA calculated from ICR-FT/MS spectra

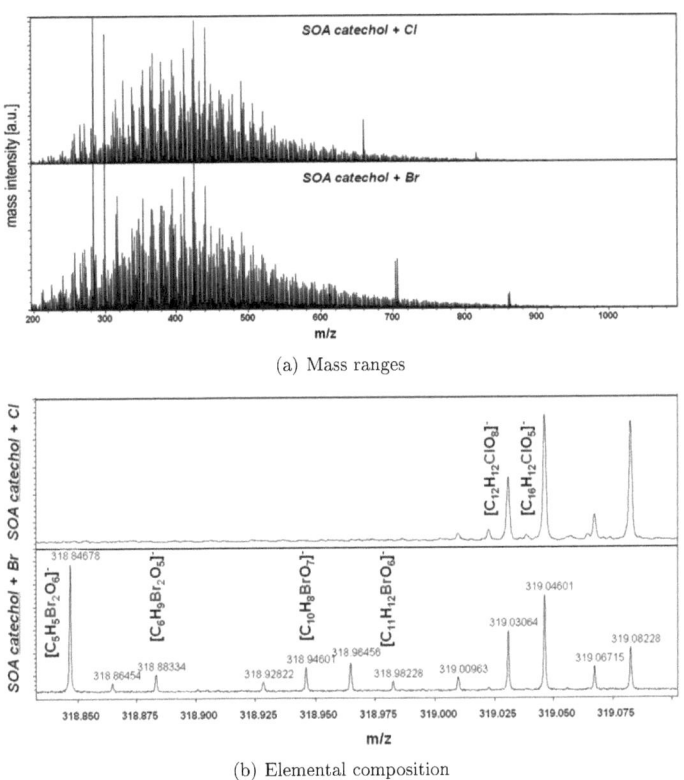

(a) Mass ranges

(b) Elemental composition

Figure 4.9: Example of mass ranges and of elemental composition of SOA from catechol processed with halogens

4.7 Changes in the aerosol optical properties

The optical properties in the UV/VIS spectral range are strongly influenced by the reaction with halogens. The diffuse-reflectance UV/VIS spectra of organic aerosols from the aromatic precursors catechol and guaiacol, reported in chapter 3, reveal strong changes in the absorption maxima resulting from the reaction with halogens (see figure 4.10).

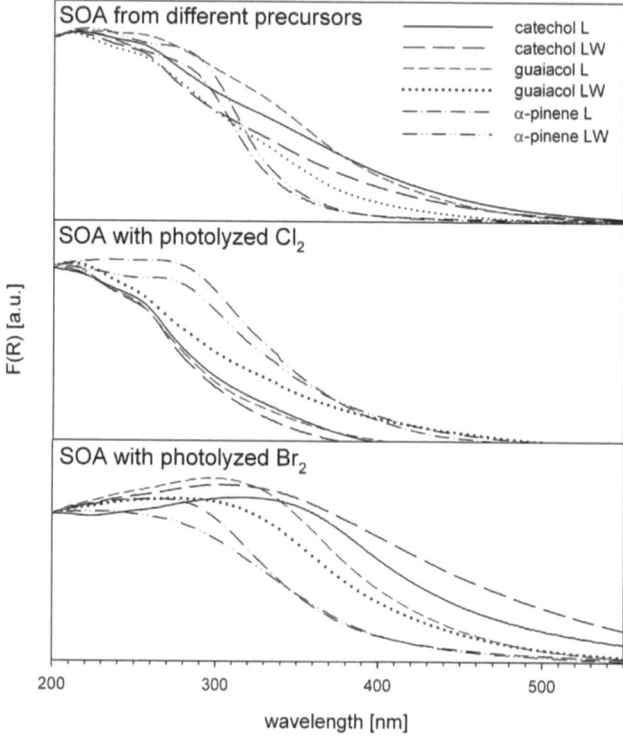

Figure 4.10: Changes in the optical properties in the UV/VIS spectral range of organic aerosols due to the reaction with gaseous halogens (L: SOA formed at 0 % relative humidity und simulated sunlight; LW: SOA formed at 25 % relative humidity and simulated sunlight)

The main absorptions at 222 and 275 nm of the unprocessed SOA shift to 210 and 250 nm after the reaction with chlorine species. The broad absorption below 600 nm is lowered in intensity and shifted to wavelengths below 400–450 nm. The effect of bromine on the diffuse-reflectance UV/VIS spectrum is totally different. While the main absorption of unprocessed

4.7 Changes in the aerosol optical properties

SOA and SOA reacted with chlorine species is located at 220 or 210 nm, respectively, SOA reacted with bromine species exhibits the maximum of absorption between 300 and 350 nm. Furthermore, the overall absorption intensity in the visible spectrum is significantly enhanced for long wavelengths.

SOA from α-pinene does not exhibit these strong influences. The reaction with halogens only results in small changes in the diffuse-reflectance UV/VIS spectrum. The absorptions at long wavelengths is slightly enhanced after reaction with bromine.

By calculating a differential absorbance, changes in the optical absorbance are emphasized (see figure 4.11). After normalizing the single absorbance spectra to an absorbance of 1 at a wavelength of 200 nm, the differential absorbance dA was calculated according to equation 4.1, where $F(R)_n^*$ is the normalized absorption spectrum (according to the Kubelka-Munk theory) of the aerosol after the reaction with the halogen, and $F(R)_n$ is the normalized absorption spectrum (also according to Kubleka-Munk) of the aerosol before the reaction.

$$dA = F(R)_n^* - F(R)_n = -lg\frac{I^*}{I_0} + lg\frac{I}{I_0} = lg\frac{I}{I^*} \qquad (4.1)$$

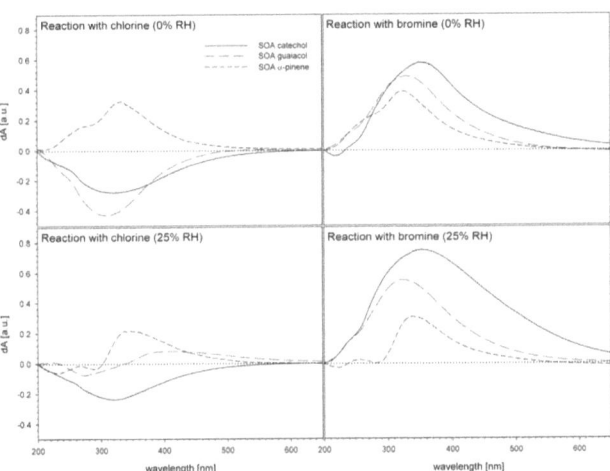

Figure 4.11: Changes in the differential absorbance dA due to the reaction of the organic aerosols with chlorine and bromine

The differential absorbance clearly indicates changes according to the electronic structure caused by the heterogeneous reaction of organic aerosols with halogens (see figure 4.11). The aromatic aerosols exhibit a negative dA at about 300 nm for the reaction with chlorine, which

4 *Molecular processing*

can be related to a "bleaching" of the aerosol. The highest maximum of absorption is shifted towards the red end of the UV/VIS spectrum. The more aliphatic SOA from α-pinene does not exhibit this behavior. Some compounds not yet identified are formed by the reaction with chlorine, which show absorbance in the region up to 350 nm.

The reaction with bromine shifts the absorption of all three aerosols up to 30 nm towards the red end of the spectrum. This behavior is especially pronounced for the aromatic SOA from catechol and least pronounced for the aliphatic SOA from α-pinene. The redshift is also observable for the filter samples. SOA processed with bromine exhibits a reddish brown to golden color.

The UV/VIS absorption of aromatic aerosols is more strongly influenced by halogens than the absorption of aliphatic aerosol from α-pinene. Hence, the interaction of atmospheric HULIS with halogen species significantly changes the absorption properties and, as a result, the interaction with solar radiation.

5 Simulated natural processing

The well-defined SOA models reported in chapter 3 were exposed to halogens generated by simulated natural halogen-release mechanisms. Two different setups were chosen for these investigations: halogens released from a simulated salt pan and halogens released from sea-salt aerosol.

Due to the rather low SOA concentrations and the high concentrations of other matrix compounds (e.g. salt from the sea-salt aerosol) the physicochemical methods described in chapter 2 are not applicable.

The two halogen release mechanisms are described in detail in section 1.5.1. Long-path absorption FTIR spectroscopy could not be used because there is no White cell installed in the Teflon chamber, and because the relative humidity used in the sea-salt aerosol experiments was too high.

5.1 Processing with halogens released from a simulated salt pan

5.1.1 Experimental setup

For the salt-pan experiments, the SOA precursors were introduced into the Teflon chamber using the impinger method. Inside the Teflon chamber, a salt pan was installed by N. Balzer and S. Bleicher from the Atmospheric Chemistry Research Laboratory. A mixture of 100 g sodium chloride and 33 mg sodium bromide with a relative humidity of 60 % was distributed on the pan and served as a substrate for the halogen release.

SOA from catechol and α-pinene was formed in the dark for 30 minutes, before the solar simulator was switched on and halogen release takes place. SOA from guaiacol had to be formed in the presence of simulated sunlight because of a lack of reactivity towards ozone without simulated sunlight. To generate the organic aerosols, 300 ppb of the precursors were

5 Simulated natural processing

used. These low concentrations, compared to the experiments described above, were necessary to avoid disturbance of the installed DOAS system.

For TPP-MS, ICR-FT/MS, and diffuse-reflectance UV/VIS spectroscopy, filter samples were employed. The ATR crystal used in the ATR-FTIR spectroscopic measurements was coated by employing the electrostatic precipitator after one hour of processing of SOA with the released halogens.

5.1.2 Aerosol size distribution

All aerosol size distributions derived from the salt-pan experiments exhibit a similar behavior of the SOA, as described in section 4.2. A slight increase in the mean particle diameters of the organic aerosols can be observed after about 60–100 minutes (Figure 5.1).

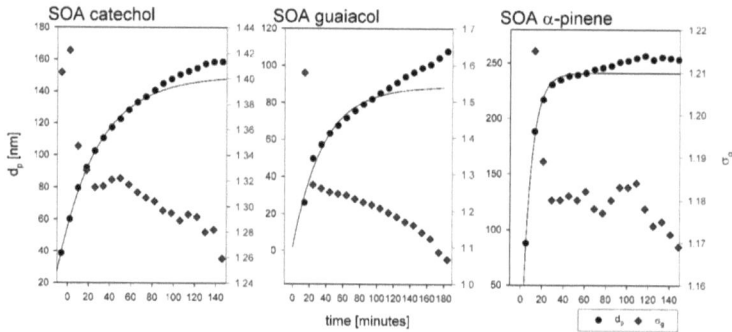

Figure 5.1: Evolution of the mean particle diameters of the organic aerosols during processing with halogens released from a simulated salt pan. The solid line indicates the expected evolution of the particle diameter without any halogen interaction.

This effect is most pronounced for SOA from guaiacol, although no formation of a second particle mode during the interaction could be observed. SOA from α-pinene exhibits first an increase of the mean particle diameter, and after 30 minutes a sudden decrease. While the root mean square deviation decreases, the mean particle diameter increases to values above the expected ones, which have been obtained by extrapolation.

These changes of the aerosol size distributions of the different model SOA are in good agreement with the observed changes during the molecular halogenation studies (see section 4.2).

5.1.3 ATR-FTIR spectroscopy

Caused by the low precursor concentrations which had to be used to avoid disturbance of the DOAS system, only few particles could be sampled onto the ATR crystal using the method of electrostatic precipitation (see section 2.5.3). The resulting ATR-FTIR spectra are shown in figure 5.2.

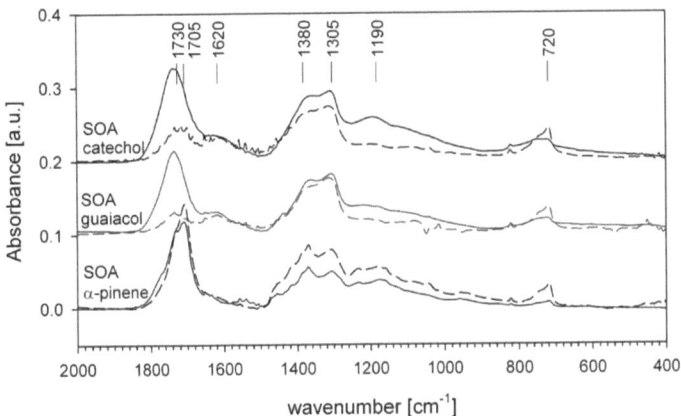

Figure 5.2: ATR-FTIR spectra of SOA processed with halogens released from a simulated salt pan: unprocessed SOA (solid lines), processed SOA (dashed lines).

For the molecular processing studies (chapter 4), a strong dominance of two absorptions at about 1380 and 1305 cm^{-1} is visible in the vibrational spectra. The absorption of the carbonyl $\nu(C=O)$ vibration decreases for SOA from catechol and guaiacol. Only SOA from α-pinene exhibits a strong $\nu(C=O)$ at about 1705 cm^{-1}.

All three SOA exhibit a rather medium to strong, broad absorption at about 720 cm^{-1} and above. This absorption can be assigned to the $\nu(C-Cl)$ vibration and indicates a possible halogenation of the organic aerosol. The shape of this absorption and the fact that it has exactly the same position for all different organic aerosols can be explained by a halogenation process involving only one specific halogen species reacting with a structural or functional feature present in all three different SOAs (e.g. carboxylic acids).

Concluding from the observations concerning the molecular processing experiments, a reaction of carboxylic groups and the subsequent formation of carbon-chlorine bonds is likely. The formation of carbon-bromine bonds, which would result in absorptions between 600 and 650 cm^{-1}, was not observed.

5.1.4 Optical properties

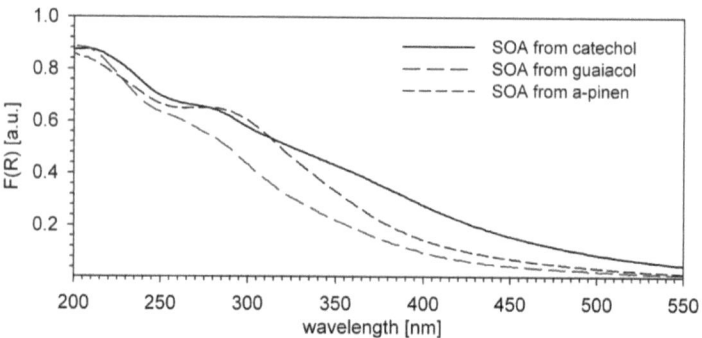

Figure 5.3: UV/VIS spectra of SOA processed with halogens released from the simulated salt pan

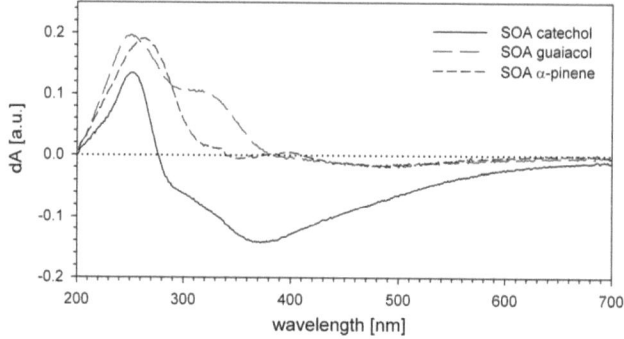

Figure 5.4: Changes in the differential absorbance dA due to the reaction of the organic aerosols with halogens released from the simulated salt pan

SOA processed by halogens released from the simulated salt pan (Figure 5.3) exhibits similar UV/VIS spectra compared with the spectra derived from SOA processed by molecular halogen species (see section 4.7).

To achieve detailed changes in the diffuse UV/VIS absorption spectra, the differential absorbance was calculated according to equation 4.1. All three SOA samples exhibit both features observed for the molecular reaction with chlorine and bromine (Figure 5.4). An increase of the differential absorbance is observed for all three different model aerosols in the spectral range around 250 nm. A decrease in dA is observed above 380 nm for all three samples, which is

5.1 Processing with halogens released from a simulated salt pan

strongest for the processed SOA from catechol. Hence, a bleaching is observed for all samples, which is strongest for the aromatic SOA from catechol.

5.1.5 Mass spectroscopy

While the assignment of thermal pyrolysis spectra to halogenated compounds was definite for SOA processed with molecular halogens, the interpretation of the temperature-programmed pyrolysis mass spectra of SOA processed with halogens released from the salt pan is less definite. This is caused by the insufficient concentration of organic aerosols in the Teflon chamber. Therefore, the identification of the isotopic ratios is not as certain as for the molecular processed organic aerosols (see section 4.5).

Strong peaks for the ratio $^{37}M/^{35}M$ of the masses are observed for all organic aerosols (Figure 5.5). These peaks do not correlate with chlorine released from the samples because the related isotopic ratio is wrong. Although the recording of the TPP-MS spectrum of SOA from catechol was disturbed by a rapid release of gaseous species at about 520 °C and a corresponding pressure increase, the increase in the $^{37}M/^{35}M$ ratio is visible at the beginning of this process.

The calculated chlorine ratio of 0.33 can only be found at low temperatures. SOA from catechol and guaiacol exhibit values close to the expected ratio between 150 and 400 °C. A more temperature-stable peak at about 400 °C is found for SOA from α-pinene. More weakly bound bromine species are observed in all three TPP-MS spectra below 200 °C.

The acquisition of ultra-high-resolution mass spectra of SOA processed with halogens released from the salt pan was impeded by matrix effects and the low organic aerosol content of the samples. Thus, only the calculated percentage of $CHOCl$- and $CHOBr$-containing species relative to the CHO-containing species is presented in table 5.1. Due to the fact that not the total mass range was evaluated, the presented amounts of halogenated compounds in table 5.1 give only hints at the halogenation processes which have taken place. Nevertheless, organic aerosols interacting with gaseous halogen species released from the salt pan contain high amounts of halogenated organic compounds. In contrast to the molecular halogen experiments, chlorination seems to be more important than bromination. Up to 52 % of chlorinated compounds in relation to 100 % of unhalogenated compounds could be measured in organic aerosol from guaiacol.

5 Simulated natural processing

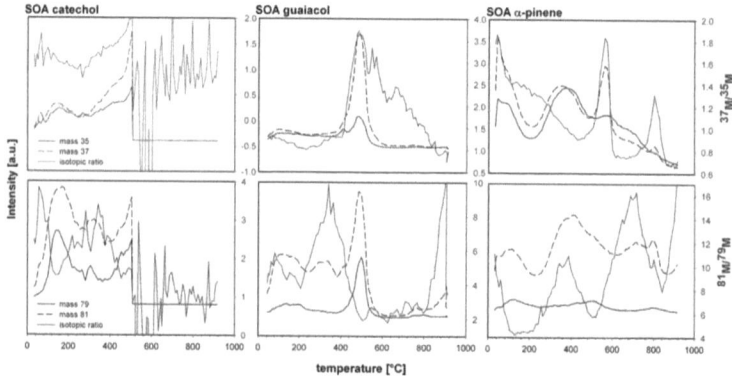

Figure 5.5: TPP-MS spectra of the SOA processed by halogens released from the salt pan: masses 35, 37, 79 and 81, representing ^{35}Cl, ^{37}Cl, ^{79}Br, and ^{81}Br, and the calculated mass ratio to determine thermal regions were the expected isotopic ratios can be found

Table 5.1: Percentage of halogenated compounds of SOA processed with halogens released from the salt pan

Elemental composition of molecules	SOA catechol	SOA guaiacol	SOA α-pinene
CHO	100 %	100 %	100 %
CHOCl	46 %	52 %	32 %
CHOBr	23 %	23 %	40 %

5.1.6 Influence of organic aerosols on halogen release mechanisms

While organic aerosols are strongly influenced by halogens released from the simulated salt pan, the halogen release mechanism itself is also strongly influenced. The modification of the ozone depletion and BrO formation is demonstrated for SOA from catechol in figure 5.6.

While the mechanism of halogen release from the salt pan causes a rapid ozone depletion and corresponding BrO formation, the organic aerosols inhibit this process. Ozone depletion is slowed down, and only less than 10 % of BrO of the expected BrO is observed. While normally also chlorine is observed (data not presented), no free chlorine could be observed in the presence of organic aerosols.

5.2 Processing with halogens released from simulated sea-salt aerosol

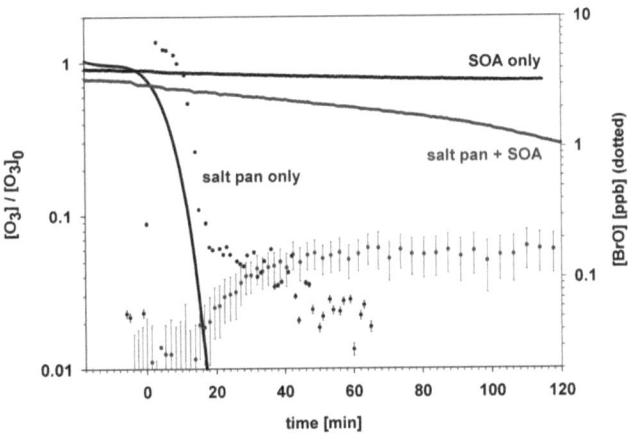

Figure 5.6: Influence of SOA from catechol on the mechanism of halogen release from the salt pan (with kind permission of N. Balzer and J. Buxmann)

5.2 Processing with halogens released from simulated sea-salt aerosol

5.2.1 Experimental setup

For the sea-salt aerosol experiments, the SOA precursors were introduced into the 700 L aerosol smog chamber using the impinger method. Aerosol was preformed at 0 % relative humidity, under simulated sunlight and using ozone. After 30 minutes, the solar simulator was switched off and simulated sea-salt aerosol was added to the smog chamber by using the internal nebulizer described in section 2.9. A solution of 5 gL^{-1} sodium chloride and 25 mgL^{-1} sodium bromide in double-distilled water was used to generate the simulated sea-salt aerosol according to Siekmann (2008). This solution was nebulized until a relative humidity above 80 % was measured in the 700 L smog-chamber. The solar simulator was switched on again when the nebulization of the sea-salt aerosol was completed.

All experiments were performed at two different SOA precursor concentrations of 1 ppm and 300 ppb. After one hour of processing of SOA by halogens released from the sea salt aerosol, filters samples were taken for TPP-MS, ICR-FT/MS and diffuse-reflectance UV/VIS spectroscopy, and the ATR crystal was coated.

5 Simulated natural processing

5.2.2 Aerosol size distribution

All organic aerosols are characterized by the appearance of two particle modes, one from the organic aerosols and a second one from the sea-salt aerosol. At the beginning of the experiments, the size of the organic aerosol particles is in the typical range of 100 to 120 nm. The inorganic salt aerosol added later exhibits much larger particle diameters between 500 and 600 nm (Figure 5.7).

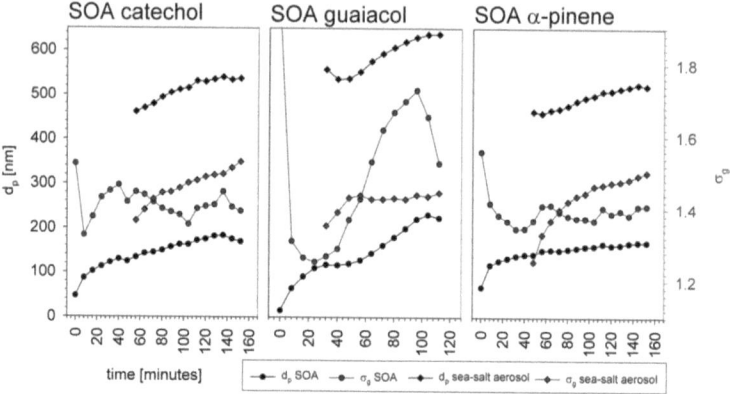

Figure 5.7: Evolution of the mean particle diameters of the organic aerosols and the sea-salt aerosol during the halogen-SOA interaction

All organic aerosols exhibit an increase in the mean particle diameter. An increase in diameter is also observed for the inorganic salt aerosol. The increase in diameter of the salt aerosol particles is caused by an increase in relative humidity in the aerosol smog chamber. The reason for the increase in particle diameter of the organic aerosol can not be determined clearly. It is either caused by the water content or by the halogen-SOA interaction.

5.2.3 ATR-FTIR spectroscopy

Caused by the high mass of inorganic aerosol on the ATR crystal, the infrared spectra of the processed SOA exhibit only few spectral details. All spectra are dominated by very strong absorptions at 1360 and 1403 cm^{-1} (Figure 5.8).

The $\nu(C=O)$ vibration at 1720 cm^{-1} is very weak; the same is true for the $\nu(C=C)$ stretching vibration at 1605 cm^{-1} in case of the aromatic aerosols. Absorptions at 776 and 740 cm^{-1} indicate the formation of halogenated species containing chlorine. The $\nu(C-Cl)$ in the infrared spectra is a good indicator for these compounds.

5.2 Processing with halogens released from simulated sea-salt aerosol

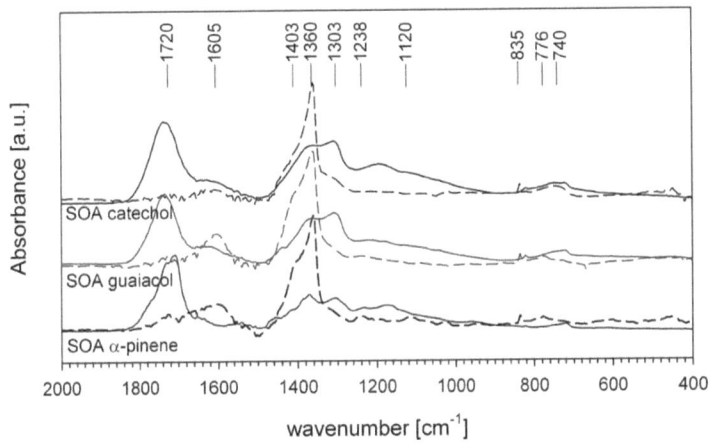

Figure 5.8: ATR-FTIR spectra of SOA processed by halogens released from the simulated sea-salt aerosol: unprocessed SOA (solid lines), processed SOA (dashed lines).

A definite assignment of the absorptions in the fingerprint region is impossible. While bands at 1360 and 1403 cm^{-1} are reported for the molecular experiments and even the salt-pan experiments, the shape of the absorption at 1360 cm^{-1} is very unusual. Further, no absorption was observed at 1403 cm^{-1} in other studies. A detailed characterization of this band is not possible based on the limited data available.

5.2.4 Optical properties

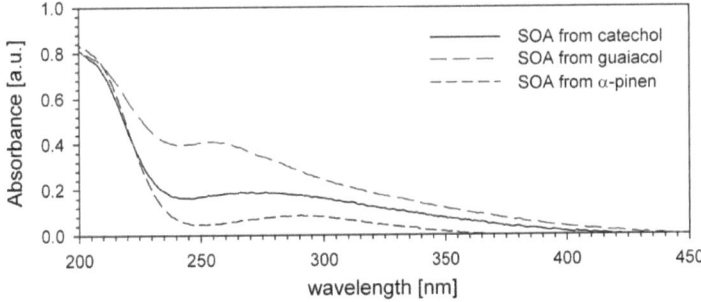

Figure 5.9: Diffuse-reflectance UV/VIS spectra of the three different model aerosols after reaction with halogens from the sea-salt aerosol

5 Simulated natural processing

Sodium chloride strongly contributes to the UV/VIS spectra by increasing the absorption close to 200 nm. Hence, the relative absorbances of the spectra are hardly comparable and thus a calculation of a differential absorbance dA does not make sense. The diffuse-reflectance UV/VIS absorption spectra are shown in figure 5.9.

5.2.5 Mass spectroscopy

The temperature-programmed pyrolysis mass spectra are disturbed by the high inorganic salt content. All TPP-MS spectra exhibit strong peaks at about 620 °C for $NaBr$. Other peaks, which would allow the calculation of isotopic ratios, are not visible (Figure 5.10).

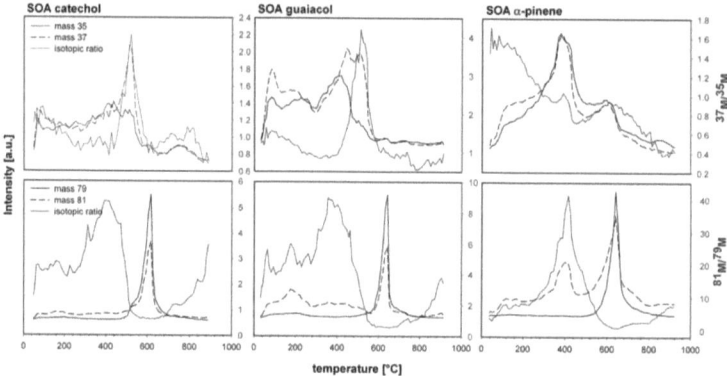

Figure 5.10: Temperature-programmed pyrolysis mass spectra of the SOA-sea-salt interaction

The ultra-high-resolution mass spectra exhibit strong matrix effects and therefore a weak signal-to-noise ratio. The percentage of halogenated species compared to the not halogenated species is shown in table 5.2. In general, the formation of chlorinated compounds in the aerosol particulate matter is more important than the formation of brominated species.

Table 5.2: Percentage of halogenated compounds of SOA processed with halogens released from sea-salt aerosol

Elemental composition of molecules	SOA catechol	SOA guaiacol	SOA α-pinene
CHO	100 %	100 %	100 %
CHOCl	72-78 %	73-94 %	73-89 %
CHOBr	21-25 %	27-94 %	15-20 %

6 Discussion and conclusions

6.1 Characterization of SOA models

The detailed physicochemical characterization of the chosen model aerosols is a prerequisite to understand the complex influences due to the heterogeneous reaction with halogen species, especially for the smog-chamber experiments, where an in-situ study of the changes in physicochemical properties of the aerosol is often hampered by a lack of instrumentation, available methods, or matrix effects.

6.1.1 SOA formation and characterization

The aliphatic SOA from α-pinene was chosen for two reasons. Particle formation from α-pinene oxidation has already been studied and reported in the literature. Furthermore, a characterization of the aerosol in the smog chamber (700 L glass chamber, see section 2.1.1) was possible by comparing reported results with obtained data, as explained in section 3.2.

The SOA formation from α-pinene is characterized by a facile preparation in the aerosol smog chamber due to the high volatility of the precursor, and the very fast particle formation process. On the other hand, this fast formation process causes methodical and experimental problems. On-line spectroscopic methods, like long-path absorption FTIR spectroscopy (see section 2.5.1), or measuring the evolution of the aerosol size distribution (see section 2.3), fail because the formation process is faster than the temporal resolution of the chosen methods. Hence, the evolution of the nucleation mode or the carbonyl stretching region could not be observed without the aid of other methods that allow a higher temporal resolution (e.g. aerosol flow reactors, see section 2.2 and 3.2.3).

Another problem is caused by the high volatility of the precursor, resulting in a comparatively high volatility of the components of the aerosol particulate matter. Thus, vacuum-dependent methods, which are used to characterize functional groups, fail (see section 3.2.4). The TPP-MS method is suitable for macromolecular aerosols with a comparatively low volatility, and it therefore fails for SOA from α-pinene.

6 Discussion and conclusions

The very high yields of SOA from α-pinene allow to sample high amounts of particulate matter for off-line methods.

Under the chosen experimental conditions, SOA from α-pinene does not exhibit strong changes. Only few changes had been observed in the ATR spectra (section 3.2.4) and the diffuse UV/VIS spectra (section 3.2.5).

In contrast to the aliphatic α-pinene, catechol and guaiacol were chosen as precursors for aromatic aerosols. This combination of three model aerosols, which differ in their structural elements and therefore in the evolution of functional groups and macromolecular structure, allows to assign the influence of halogen species on different features.

SOA from catechol or guaiacol typically exhibits small particles at the chosen experimental conditions, with diameters between 40 and 90 nm, assembled in a very fast formation process. The yields, though, seem to be much lower than observed for SOA from α-pinene.

Environmental conditions like solar radiation or relative humidity influence SOA from catechol and guaiacol more strongly than SOA from α-pinene. Not only physical properties like aerosol size distributions or formation yields are changed, but also chemical properties like the total amount of oxidized sites as well as the amount and types of functional groups. The presence of simulated sunlight or variation of relative humidity results in various amounts of degrading structural elements, as revealed by different intensities of aromatic $\nu(C = C)$ and $\nu(C - H)$ stretching vibrations, and pronounced oxygen-containing functional groups, which shifts the position of $\nu(C = O)$ vibrations.

Light absorption of those organic particles ranges up to 600 nm into the visible region and declines very smoothly. Thus, derived SOA samples are colored light brown. This optical feature indicates absorption processes related to conjugated structures with a large variety of chemically bound oxygen.

The chemical transformation from the gaseous precursor to the final aerosol particle is characterized by the formation of different functional groups and the disappearance of well-defined structural elements of the entire benzene ring. The aromatic or olefinic structural element still persists in the aerosol particle.

Therefore, SOA from the aromatic precursors catechol and guaiacol exhibits, in contrast to SOA from α-pinene, properties like a high absorbance in the visible spectral range, an aromatic or olefinic structure, and the formation of highly oxidized functional groups. Due to these properties, SOA from catechol and guaiacol might serve as a model aerosol for atmospheric HULIS, as described in detail in the following section.

6.1.2 HULIS-like behavior of SOA from catechol and guaiacol

SOA from catechol and guaiacol provide several features similar to natural and synthetic HULIS and commercial proxies, as discussed above. Especially the high molecular weight, caused by the aromatic system, and the polycarboxylic acidic functionality matches those properties. Also the most significant reported group frequencies are well comparable to those of humic acids, natural HULIS, and other model HULIS.

The difference between the observed H/C values and the reported values for atmospheric HULIS is explained by the absence of aliphatic side chains in the SOA from catechol and guaiacol. Thus, SOA from these precursors represent the aromatic and olefinic oxidized core structure of atmospheric HULIS very well. Only saturated aliphatic parts of atmospheric HULIS are not represented. SOA from catechol or guaiacol is located in the $n_C/\overline{OS_C}$-diagram close to WSOC, but also exhibits higher carbon oxidation states (see figure 6.1). This type of SOA is characterized by a high degree of functionalization and oligomerization.

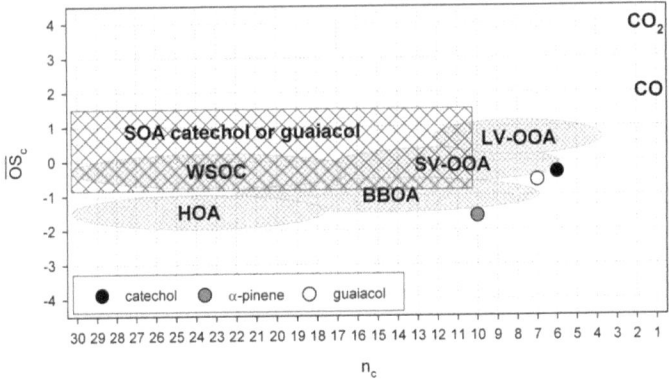

Figure 6.1: Classification of SOA from catechol or guaiacol in the $n_C/\overline{OS_C}$-diagram

Especially infrared-spectroscopic results confirm this view. ATR-FTIR spectra of SOA from catechol and guaiacol exhibit absorptions reported for natural HULIS and accepted proxies like SRFA. Long-path FTIR spectroscopy and aerosol flow-reactor studies characterized the evolution of the carbonyl stretching region, which appears similar to that reported from HULIS.

The carbonyl and aromatic stretching regions between 1600 and 1800 cm^{-1} are dominated by two bands at 1620 and 1716 or 1740. These absorptions are reported for SRFA and for natural HULIS by Salma et al. (2010). Especially the carbonyl stretching vibration at 1716 cm^{-1} observable for SOA from catechol formed in the dark was found in BBOA by these authors.

6 Discussion and conclusions

The most important functional groups of SOA from catechol and guaiacol seem to be carboxylic acids, carboxylic anhydrides, and lactones or esters, indicated by ATR absorptions at 1317 and 1295 cm^{-1}. The presence of these groups is confirmed by temperature-programmed pyrolysis mass spectroscopy. These groups have been reported for a natural aerosol at a rural site with an aromatic content of 17 % (Coury and Dillner, 2009). Samburova et al. (2007) report high carboxylic and arylic fractions in HULIS samples. Aromatic acids with high molecular weights were found by Stone et al. (2009) to correlate well with HULIS. Carboxylic acids and their modifications are commonly found in natural HULIS, biomass burning aerosol, and water-soluble organic carbon (WSOC) (Limbeck and Puxbaum, 1999; Salma and Lang, 2008; Kundu et al., 2010; Kumagai et al., 2010).

Sun et al. (2010) report dimer formation based on $C-O$ or $C-C$ bond formation. The ATR-FTIR spectra of SOA from catechol and guaiacol measured in this study exhibit also hints at the formation of carboxylic acid dimers.

In principle, all spectra of natural HULIS look rather similar, which is not surprising since these are mixtures of highly oxidized organic compounds. Small differences in band positions and relative intensities are related to the particular sampling site and the history of the aerosol.

The main group frequencies of SOA ($\nu(O-H)$, $\nu(C=O)$, $\nu(C=C)_{aromatic}$, $\nu(C-O)$ of carboxylic acids and of ethers) from catechol and guaiacol are compared with natural HULIS extracts (Polidori et al., 2008; Duarte et al., 2005), SRFA (Graber and Rudich, 2006), fulvic acids from soil samples (Stevenson and Goh, 1971), humic and fulvic acids from peat samples(Baes and Bloom, 1989), as well as tannic acid, which is a previously reported model compound (Cowen and Al-Abadleh, 2009).

In general, the assignment of the main characteristic group frequencies was done in the same way as reported in literature (see table 6.1). Only few discrepancies occurred during the assignment, which are marked in the table. However, in two studies (Polidori et al., 2008; Stevenson and Goh, 1971) the absorption at about 1620 cm^{-1}, which is also present in all spectra, has been interpreted as an additional $\nu(C=O)$ absorption. Based on the reported long-path absorption spectra of SOA from catechol and guaiacol, and according to Graber and Rudich (2006) and Cowen and Al-Abadleh (2009), an assignment to $\nu(C=C)$ seems to be the more comprehensible interpretation. Infrared bands of amides, which were observed in the spectra of some field samples, can not be found in our lab studies, since nitrogen-containing compounds are absent, which was confirmed by ultra-high-resolution mass spectroscopy. While the main absorptions are in excellent agreement with the literature, differences among all five studies occur only in the assignment of $\nu(C-O)$ of the carboxylic acids. The reason for this might be the different degree of dissociation of the carboxylic acids or a potential hydrogen bonding, depending on the environmental conditions.

SOA from catechol or guaiacol can be prepared in an aerosol smog chamber very easily. Other model compounds for atmospheric HULIS need complex preparation techniques. Commercial HULIS proxies, like SRFA, can only be used by ultrasonically nebulizing or atomizing their solutions. Synthetic SOA from catechol and guaiacol produced in aerosol smog-chamber experiments meet the physicochemical characteristics of HULIS better than SRFA according to the analytical results, and can be used as atmospheric model substances for HULIS in laboratory experiments, except for the missing aliphatic side chains.

6 Discussion and conclusions

Table 6.1: Comparison of infrared spectral features of natural humic acids, HULIS, aerosol extracts and two SOA models (w- weak; m- medium; s -strong)

Assignments [cm^{-1}]	Natural humic and fulvic acids[1]	Natural humic and fulvic acids[2]	Extracts from rural aerosol[3]	Extracts from urban aerosol[4]	Tannic acid model[5]	aromatic SOA (present study)
$\nu(O-H)_{aromatic}$	3400	3300-3400	3400 m	3360-3444 m	3392-3400 m	3330-3450 m
$\nu(C=O)$	1720 m	1720	1720 s	1710-1742 s	1713 m	1716-1740 s
$\nu(C=C)$	1600-1660	1620	1600-1660 m	1613-1638 m[a]	1616 m	1620 m
$\nu(C-O)$ of $-COOH$	1200	1225,1350	1220	1439,1375[b]	1209,1298,1250-1310	1295,1317,1416
$\nu(C-O)$ of ethers	1050[c]		1061[c]	1045,1090,1177,1217w	1036,1045,1080w	1045,1118w

[1] (Stevenson and Goh, 1971)
[2] (Baes and Bloom, 1989)
[3] (Graber and Rudich, 2006)
[4] (Polidori et al., 2008)
[5] (Cowen and Al-Abadleh, 2009)
[a] interpreted as $\nu(C=O)$
[b] assigned to $-COO-$
[c] assigned to polysaccharides

6.2 Halogenation of SOA using molecular halogen species

Studying the heterogeneous reaction of the well-defined model aerosols with molecular halogens photolyzed by the UV/VIS irradiation of the solar simulator gave access to the understanding of specific pathways of transformation with respect to the physicochemical properties.

In general, halogens influence several properties of organic aerosols, but different SOA exhibit different transformations. The reactions are depending on the SOA precursor and thus on the chemical composition of the organic aerosol, but also on the halogen species and also the experimental conditions, which can even change the predominant halogenated species (da Rosa and Zetzsch, 2001).

Organic aerosols are strongly influenced with respect to their aerosol size distributions. Chlorine species cause a strong increase of the mean particle diameters, and in the case of weakly bonded ethers, the formation of a second mode has been observed. The term aerosol must in this case be used according to its exact definition: particles and their surrounding gas phase, which contains, in addition to the common atmospheric gas-phase species, also volatile oxidation products of the aerosol formation process. This is especially important for aerosol smog-chamber studies, where particles and gas phase can be separated from meteorological effects. Therefore, the increase in particle diameter can be explained by the formation of low-volatile species from the surrounding gas phase, condensing onto the already present particles. Due to this effect, atmospheric halogen species might also exhibit an aerosol-formation potential as soon as the oxidation capacity of the atmosphere is not high enough to produce further low-volatile compounds. The effect of bromine species on the aerosol size distributions is weaker than observed for chlorine. On the other hand, bromine is able to reduce the mean particle diameter, as observed for SOA from α-pinene. This effect can be explained by halogen species not only forming new low-volatile compounds, thus contributing to existing particles or forming new particles, but also changing the chemical structure of the particulate matter itself.

The last statement is underlined by the observation of the formation of various gaseous species due to the interaction of halogens with organic aerosols. Long-path absorption spectra indicate the formation of additional amounts of CO_2, CO, and HCl or HBr. At the same time, a decrease in $\nu(C-H)_{aromatic}$ and $\nu(C-H)_{aliphatic}$ of the particle phase was observed. Thus, $C-H$ bonds are destroyed. This is known to be the case for the reaction of chlorine with aliphatic compounds, but also bromine species show this behavior. Therefore, the observations might be explained by the reaction of BrO or $HOBr$ with $C-H$ bonds.

The spectra of the particulate matter of the organic aerosols itself also exhibit strong changes. While the carbonyl $\nu(C=O)$ is the most dominant vibration of the unprocessed aerosols, this vibration is less important after the reaction with halogens. This decrease can also be observed

6 Discussion and conclusions

for the $\nu(C=C)$ of the aromatic aerosols. The $\nu(C-Cl)$ and the $\nu(C-Br)$ stretching vibrations indicate the presence of halogenated compounds, even though the assignment of the vibrations to specific compounds was not possible. Also $\nu(O-Cl)$ (Evans et al., 1965) and spectral features of structural elements (Socrates, 1980) can be found in this spectral region. Most dominant are bands observed at about 1370 and 1300 cm^{-1}. The assignment of those absorptions is difficult. Based on the observation that these bands are observed for halogenation processes with chlorine as well as with bromine at the same position, the influence of a functional group or a structural element is more likely than the formation of a halogenated compound. Therefore, a link of these absorptions with carboxylic acids, carboxylic acid salts or derivates is likely but not confirmed.

The transformations of the particulate matter also affect the optical properties of the aerosols in the UV/VIS spectral range. While the reaction with chlorine causes a bleaching of the particles, the reaction with bromine causes a higher absorbance towards higher wavelengths – the particulate phase appears "golden".

The presence of halogenated compounds in the particulate matter of the organic aerosols is confirmed by mass spectroscopic methods used in this study. High amounts of brominated and chlorinated compounds could be identified using ultra-high-resolution mass spectroscopy.

Based on the fact that the particle diameter increases during the reaction of SOA with halogens, which appears to be coupled to a decrease of the saturation vapor pressure, halogenated SOA is located at low vapor pressures and over a wide range of O/C ratios in the O/C-diagram (Figure 6.2(a)). Furthermore, the processed organic aerosol exhibits high carbon oxidation states and a wide range of carbon atoms per macromolecule (Figure 6.2(b)).

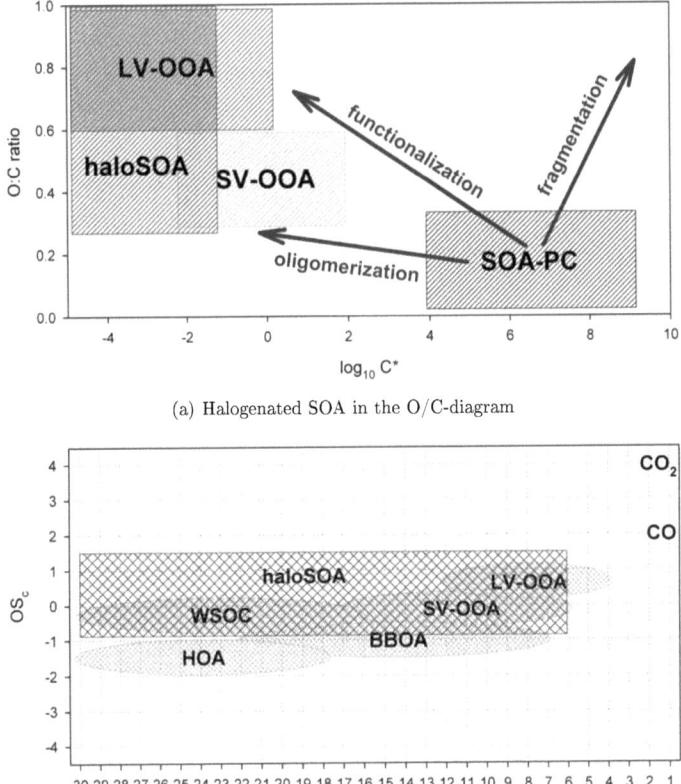

(a) Halogenated SOA in the O/C-diagram

(b) Halogenated SOA in the $n_C/\overline{OS_C}$-diagram

Figure 6.2: Halogenated SOA in different currently used diagrams

6 Discussion and conclusions

6.3 Comparison of the results of natural halogenation processes to the molecular processing study

To obtain details on possible natural halogenation processes involving organic aerosols, the model aerosols were exposed to natural systems which are known to release gaseous halogen species. The complex data obtained from these systems exhibit strong matrix effects, which hamper the detailed interpretation of the data sets. Thus, the results of the molecular processing studies were used to identify physicochemical changes related to the organic aerosols.

Results from the interaction of organic aerosols with halogens released from the simulated salt pan can be compared quite well to the observed molecular processing. A similar increase of the particle diameters is observed. ATR infrared spectra exhibit the same spectral features. The optical properties in the UV/VIS spectral range are comparable to the observed spectra from the molecular experiments and redshifts as well as blueshifts are both visible. Although the mass spectra exhibit a poor signal-to-noise ratio, the formation of halogenated compounds seems to be confirmed for the particulate matter.

The interpretation of results of the sea-salt aerosol interaction experiments is more difficult. The increase in mean diameter in the aerosol size distribution is also observed for the organic aerosols, but this increase is observed for the sea-salt aerosol as well, and might also be related to the very high water content of the system. At detailed interpretation of the optical spectra is hardly possible due to the strong influence of the salt matrix. Statements on the degree of halogenation based on mass spectra are more difficult than for the salt-pan aerosol, because the collected filter samples contain also halogens from the salt matrix which are present as an aerosol. For this system, the halogenation process of the organic aerosols could not be determined.

In general, the investigation of the transformation of physicochemical features of the organic aerosols led to the first steps of an understanding of the complex processes involved. A detailed understanding was not achieved at the present time, as it would require more molecular-based techniques to analyze the processes.

6.4 Influence of natural halogen species on organic aerosols

A strong influence of atmospheric halogen species on organic aerosols is indicated by the performed aerosol smog-chamber experiments. Several parameters used to characterize the model

6.4 Influence of natural halogen species on organic aerosols

aerosols are changed by the halogenation process. These changes are depending on the particular aerosol as well as on the halogen source. The chemical composition of the organic aerosols determines the details of the changes. Hence, no general "road map" for atmospheric halogenation processes could be identified.

A simplified scheme, summarizing possible effects of halogens on different SOA, is given in figure 6.3. Depending on the kind of SOA and the halogen source, the change in the aerosol size distribution can be observed. The chemical transformation and the related change in optical properties are indicated by the change in the color of the halogenated organic aerosols (haloSOA). The possible formation of secondary particles from the reacted SOA (SSOA) is indicated. Further, the main identified gaseous species released from the aerosol are specified.

The reaction of organic aerosols with halogens from natural sources is confirmed by the influence of organic particles on the halogen release-mechanisms themselves, as reported in section 5.1.6.

Halogens take part in the aging process of organic aerosols (Figure 6.4). Based on the results of the present study, interactions occur in the gas phase as well as in the particle phase. The gas-phase halogenation generates additional low-volatile compounds, which condense onto the existing particles. Furthermore, the halogenation process releases simple gaseous molecules. The decrease in vapor pressure can also lead to new particle formation, caused by the halogenation of the SOA precursor or the oxidized SOA precursor. Therefore, halogens can interact in ways with organic molecules (possible SOA precursors) similar to those suggested in the current literature on nucleation and aging (Kroll et al., 2011).

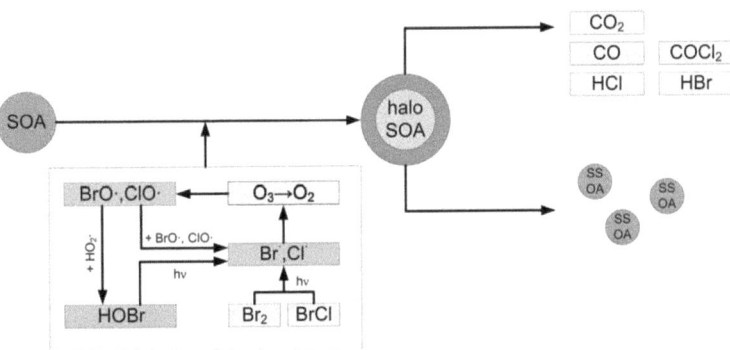

Figure 6.3: Simplified scheme summarizing possible effects of halogens on secondary organic aerosols: transformation of SOA to haloSOA and formation of gaseous species and secondary SOA (SSOA).

6 Discussion and conclusions

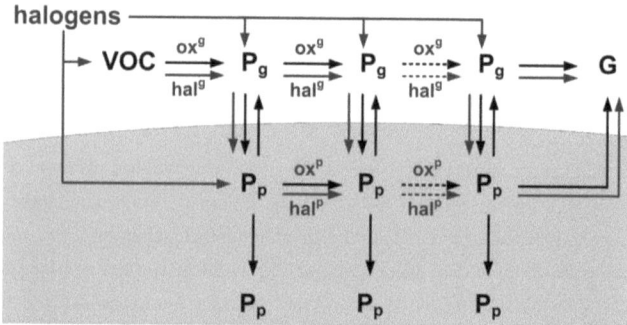

Figure 6.4: Influence of halogens on the nucleation process and aging of SOA

6.5 Applicability of aerosol smog-chamber studies and physicochemical methods

Finally, a discussion on the applicability of aerosol smog-chamber experiments for determining heterogeneous reaction schemes based on physicochemical methods is added.

While the experimental setup of an aerosol smog chamber is the most realistic approximation of environmental conditions, obtaining detailed physicochemical information on a molecular scale is very difficult or in some cases nearly impossible. A most complex chemical system, especially with regard to heterogeneous processes involving organic aerosols, is inherent according to the most realistic setup of a smog chamber. A detailed investigation of those processes is only possible by splitting the complex reaction scheme into smaller and simpler parts, and comparing the results of those model experiments with the overall process. Other experimental setups, such as aerosol flow reactors and vacuum reaction cells coupled with spectroscopic methods, are able to provide more insight into chemical processes on a molecular scale.

Apart from the difficult access to the physicochemical parameters of the reaction of interest, aerosol smog chambers pose another problem resulting from the small amount of samples which could be obtained from a smog-chamber run depending on the chosen concentrations. While these concentrations should be as realistic as possible in comparison with natural conditions, the available amount of gaseous or particulate samples is limited by this precondition. In contrast to remote sampling, the amount of sampled gaseous and particulate matter is limited by the volume of the aerosol smog chamber. Furthermore, the removal of particles by sedimentation and diffusion to the chamber walls is depending on the surface-to-volume ratio and the geometry of the chamber (for further details on this topic, see Crump and Seinfeld (1980); Crump et al.

6.5 Applicability of aerosol smog-chamber studies and physicochemical methods

(1983)). Thus, the maximum time to study aerosols in an aerosol smog chamber is limited by these parameters.

While up-to-date methods, like aerosol mass spectroscopy, have been developed to handle those low concentrations, the methods used for this study, which belong to the classical methods for the structure determination of chemical compounds, are hampered due to the fact that still a rather high amount of sample is needed to obtain the desired physicochemical information. On the other hand, these methods allow a deep insight into the chemical nature of organic aerosols and the involved heterogeneous reactions.

References

Abdel-Salam, M., Nakano, M., and Mizuno, A.: Electric fields and corona currents in needle-to-meshed plate gaps, J. Phys. D: Appl. Phys., 40, 3363–3370, 2007.

Ahn, M.-Y., Martinez, C. E., Archibald, D. D., Zimmerman, A. R., Bollag, J.-M., and Dec, J.: Transformation of catechol in the presence of a laccase and birnessite, Soil Bio. Biochem., 38, 1015–1020, 2006.

Allen, D. T., Palen, E. J., Haimov, M. I., Hering, S. V., and Young, J. R.: Fourier transform infrared spectroscopy of aerosol collected in a low pressure impactor (LPI/FTIR): method development and field calibration, Aerosol Sci. Technol., 21, 325–42, 1994.

Andreae, M. O.: A new look at aging aerosols, Science, 326, 1493–1494, 2009.

Andreae, M. O. and Crutzen, P. J.: Atmospheric aerosols: biogeochemical sources and role in atmospheric chemistry, Science, 276, 1052–1058, 1997.

Atkinson, R.: Gas-phase tropospheric chemistry of organic compounds, J. Phys. Chem. Ref. Data Monograph., 2, 1–216, 1994.

Atkinson, R.: Gas-phase tropospheric chemistry of volatile organic compounds: 1. Alkanes and alkenes, J. Phys. Chem. Ref. Data, 26, 215–290, 1997.

Baduel, C., Voisin, D., and Jaffrezo, J. L.: Comparison of analytical methods for Humic Like Substances (HULIS) measurements in atmospheric particles, Atmos. Chem. Phys., 9, 5949–5962, 2009.

Baes, A. U. and Bloom, P. R.: Diffuse Reflectance and Transmission Fourier Transform Infrared (DRIFT) spectroscopy of humic and fulvic acids, Soil Sci. Soc. Am. J., 53, 695–700, 1989.

Baltensperger, U., Kalberer, M., Dommen, J., Paulsen, D., Alfarra, M. R., Coe, H., Fisseha, R., Gascho, A., Gysel, M., Nyeki, S., Sax, M., Steinbacher, M., Prevot, A. S. H., Sjogren, S., Weingartner, E., and Zenobi, R.: Secondary organic aerosols from anthropogenic and biogenic precursors, Faraday Discuss., 130, 265–278, 2005.

References

Balzer, N., Buxmann, J., Bleicher, S., Ofner, J., Kotte, K., Zetzsch, C., and Platt, U.: Release of reactive halogen species from a simulated salt pan, using dry and wet solid NaCl/NaBr surfaces and environmental samples in smog-chamber experiments, 3. Bi-Annuales Kieler Symposium "Future Ocean", Kiel / Germany, 2010.

Becker, K. H. and Klein, T.: Atmosphärische Oxidation aromatischer Verbindungen und ausgewählter dicarbonylischer Folgeprodukte, Tech. rep., Fachbereich Physikalische Chemie - Bergische Universität Gesamthochschule Wuppertal, 1987.

Behnke, W. and Zetzsch, C.: Heterogeneous photochemical formation of Cl atoms from NaCl aerosol, NOx and ozone, J. Aerosol Sci., 21, 229–232, 1990.

Bejan, I. G.: Investigations on the gas phase atmospheric chemistry of nitrophenols and catechols, Ph.D. thesis, Bergische Universität Gesamthochschule Wuppertal, 2006.

Berndt, T., Böge, O., and Stratmann, F.: Gas-phase ozonolysis of α-pinene: gaseous products and particle formation, Atmos. Environ., 37, 3933–3945, 2003.

Bleicher, S., Balzer, N., Zetzsch, C., Buxmann, J., and Platt, U.: Kinetic smog-chamber studies on halogen activation from a simulated salt pan, using dry and wet NaCl/NaBr surfaces, EGU General Assembly 2010, Vienna / Austria, 2010.

Boyce, S. D. and Hornig, J. F.: Reaction pathways of trihalomethane formation from the halogenation of dihydroxyaromatic model compounds for humic acid, Environ Sci Technol., 17, 202–211, 1983.

Brauers, T. and Wiesen, P.: Experiments in large simulation chambers, Chem. Unserer Zeit, 41, 212–218, 2007.

Bröske, R., Kleffmann, J., and Wiesen, P.: Heterogeneous conversion of NO_2 on secondary organic aerosol surfaces: a possible source of nitrous acid (HONO) in the atmosphere?, Atmos. Chem. Phys., 3, 469–474, 2003.

Burger, T., Ploss, H. J., Kuhn, J., Ebel, S., and Fricke, J.: Diffuse reflectance and transmittance spectroscopy for the quantitative determination of scattering and absorption coefficients in quantitative powder analysis, Appl. Spectrosc., 51, 1323–1329, 1997.

Campbell, D., Copeland, S., and Cahill, T.: Measurement of aerosol absorption coefficient from teflon filters using integrating plate and integrating sphere techniques, Aerosol Sci. Technol., 22, 287–292, 2011.

Caregnato, P., Gara, P. D., Bosio, G. N., Martire, D. O., and Gonzalez, M. C.: Reactions of $Cl\cdot/Cl_2\cdot$ - radicals with the nanoparticle silica surface and with humic acids: model reactions for the aqueous phase chemistry of the atmosphere, Photochem. Photobiol., 83, 944–951, 2007.

Carpenter, L. J., Hopkins, J. R., Jones, C. E., Lewis, A. C., Parthipan, R., Wevill, D. J., Poissant, L., Pilote, M., and Constant, P.: Abiotic source of reactive organic halogens in the sub-arctic atmosphere?, Environ. Sci. Technol., 39, 8812–8816, 2005.

Carrasco, N., Rayez, M. T., Rayez, J. C., and Doussin, J. F.: Experimental and theoretical study of the reaction of OH radical with sabinene, Phys. Chem. Chem. Phys., 8, 3211–3217, 2006.

Chan, M. N. and Chan, C. K.: Hygroscopic properties of two model humic-like substances and their mixtures with inorganics of atmospheric importance, Environ. Sci. Technol., 37, 5109–5115, 2003.

Claeys, M., Graham, B., Vas, G., Wang, W., Vermeylen, R., Pashynska, V., Cafmeyer, J., Guyon, P., Andreae, M. O., Artaxo, P., and Maenhaut, W.: Formation of secondary organic aerosols through photooxidation of isoprene, Science, 303, 1173–1176, 2004.

Claeys, M., Szmigielski, R., Kourtchev, I., van der Veken, P., Vermeylen, R., Maenhaut, W., Jaoui, M., Kleindienst, T. E., Lewandowski, M., Offenberg, J. H., and Edney, E. O.: Hydroxydicarboxylic acids: Markers for secondary organic aerosol from the photooxidation of α-pinene, Environ. Sci. Technol., 41, 1628–1634, 2007.

Coeur-Tourneur, C., Tomas, A., Guilloteau, A., Henry, F., Ledoux, F., Visez, N., Riffault, V., Wenger, J. C., and Bedjanian, Y.: Aerosol formation yields from the reaction of catechol with ozone, Atmos. Environ., 43, 2360–2365, 2009.

Coury, C. and Dillner, A. M.: A method to quantify organic functional groups and inorganic compounds in ambient aerosols using attenuated total reflectance FTIR spectroscopy and multivariate chemometric techniques, Atmos. Environ., 42, 5923–5932, 2008.

Coury, C. and Dillner, A. M.: ATR-FTIR characterization of organic functional groups and inorganic ions in ambient aerosols at a rural site, Atmos. Environ., 43, 940–948, 2009.

Cowen, S. and Al-Abadleh, H. A.: DRIFTS studies on the photodegradation of tannic acid as a model for HULIS in atmospheric aerosols, Phys. Chem. Chem. Phys., 11, 7838–7847, 2009.

Crump, J. G. and Seinfeld, J. H.: Aerosol behavior in the continuous stirred tank reactor, AIChE J., 26, 610–16, 1980.

Crump, J. G., Flagan, R. C., and Seinfeld, J. H.: Particle wall loss rates in vessels, Aerosol Sci. Technol., 2, 303–9, 1983.

References

da Rosa, M. and Zetzsch, C.: Influence of pH and halides on halogen species in the aqueous phase, J. Aerosol Sci., 32, 311, 2001.

da Rosa, M. B.: Untersuchungen heterogener troposphärenrelevanter Reaktionen von Schwefel- und Halogenverbindungen, Ph.D. thesis, University of Hannover, 2003.

Dandekar, A., Baker, R. T. K., and Vannice, M. A.: Characterization of activated carbon, graphitized carbon fibers and synthetic diamond powder using TPD and DRIFTS, Carbon, 36, 1821–1831, 1998.

Davis, S. P., Abrams, M., and Brault, J.: Fourier transform spectrometry, Academic press, San Diego, California, 2001.

Dekermenjian, M., Allen, D. T., Atkinson, R., and Arey, J.: FTIR analysis of aerosol formed in the ozone oxidation of sesquiterpenes, Aerosol Sci. Technol., 30, 349–363, 1999.

Dinar, E., Taraniuk, I., Graber, E. R., Katsman, S., Moise, T., Anttila, T., Mentel, T. F., and Rudich, Y.: Cloud condensation nuclei properties of model and atmospheric HULIS, Atmos. Chem. Phys., 6, 2465–2482, 2006.

Donahue, N. M., Robinson, A. L., Stanier, C. O., and Pandis, S. N.: Coupled partitioning, dilution, and chemical aging of semivolatile organics, Environ. Sci. Technol., 40, 2635–2643, 2006.

Duarte, R. M. B. O., Pio, C. A., and Duarte, A. C.: Spectroscopic study of the water-soluble organic matter isolated from atmospheric aerosols collected under different atmospheric conditions, Anal. Chim. Acta, 530, 7–14, 2005.

Eiden, R.: Determination of the complex index of refraction of spherical aerosol particles, Appl. Opt., 10, 749–754, 1971.

Enami, S., Vecitis, C. D., Cheng, J., Hoffmann, M. R., and Colussi, A. J.: Global inorganic source of atmospheric bromine, J. Phys. Chem. A, 111, 8749–8752, 2007.

Evans, J. C., Lo, G. Y. S., and Chang, Y. L.: The vibrational spectra of acetyl hypochlorite, Spectrochim. Acta, 21, 973–9, 1965.

Excoffon, P. and Marechal, Y.: Infrared spectra of H-bonded systems: saturated carboxylic acid dimers, Spectrochim. Acta, 28A, 269–283, 1972.

Fahimi, I. J., Keppler, F., and Schöler, H. F.: Formation of chloroacetic acids from soil, humic acid and phenolic moieties, Chemosphere, 52, 513–420, 2003.

Farman, J., Gardiner, B., and Shanklin, J.: Large losses of total ozone in Antarctica reveal seasonal ClO_x/NO_x interaction, Nature, 315, 207–210, 1985.

Fierz, M., Kaegi, R., and Burtscher, H.: Theoretical and experimental evaluation of a portable electrostatic TEM sampler, Aerosol Sci. Technol., 41, 520–528, 2007.

Fine, P. M., Cass, G. R., and Simoneit, B. R. T.: Chemical characterization of fine particle emissions from the fireplace combustion of woods grown in the southern United States, Environ. Sci. Technol., 36, 1442–1451, 2002.

Fine, P. M., Cass, G. R., and Simoneit, B. R. T.: Chemical characterization of fine particle emissions from the wood stove combustion of prevalent United States tree species, Environ. Eng. Sci., 21, 705–721, 2004.

Finlayson-Pitts, B. J.: The tropospheric chemistry of sea salt: A molecular-level view of the chemistry of NaCl and NaBr, Chem. Rev., 103, 4801–4822, 2003.

Finlayson-Pitts, B. J.: Reactions at surfaces in the atmosphere: integration of experiments and theory as necessary (but not necessarily sufficient) for predicting the physical chemistry of aerosols, Phys. Chem. Chem. Phys., 11, 7760–7779, 2009.

Finlayson-Pitts, B. J.: Halogens in the troposphere, Anal. Chem., 82, 770–776, 2010.

Finlayson-Pitts, B. J. and Pitts, J. N.: Chemistry of the upper and lower atmosphere, Academic Press, San Diego London, 2000.

Fisseha, R., Dommen, J., Sax, M., Paulsen, D., Kalberer, M., Maurer, R., Höfler, F., Weingartner, E., and Baltensperger, U.: Identification of organic acids in secondary organic aerosol and the corresponding gas phase from chamber experiments, Anal. Chem., 76, 6535–6540, 2004.

Flett, M. S. C.: Intensities of some group characteristic infra-red bands, Spectrochim. Acta, 18, 1537–1556, 1962.

Florio, G. M., Zwier, T. S., Myshakin, E. M., Jordan, K. D., and Sibert III, E. L.: Theoretical modeling of the OH stretch infrared spectrum of carboxylic acid dimers based on first-principles anharmonic couplings, J. Chem. Phys., 118, 1735–1746, 2003.

Forster, P., Ramaswamy, V., Artaxo, P., Berntsen, T., Betts, R., Fahey, D., Haywood, J., Lean, J., Lowe, D., Myhre, G., Nganga, J., Prinn, R., Raga, G., M., S., and Van Dorland, R.: Changes in Atmospheric Constituents and in Radiative Forcing, in: Climate Change 2007: The Physical Science Basis. Contribution of Working Group I to the Fourth Assessment

References

Report of the Intergovernmental Panel on Climate Change, edited by Solomon, S., Qin, D., Manning, M., Chen, Z., Marquis, M., Averyt, K., Tignor, M., and Miller, H., Cambridge University Press, Cambridge, United Kingdom and New York, 2007.

Forstner, H. J. L., Flagan, R. C., and Seinfeld, J. H.: Secondary organic aerosol from the photooxidation of aromatic hydrocarbons: Molecular composition, Environ. Sci. Technol., 31, 1345–1358, 1997.

Foster, K. L., Plastridge, R. A., Bottenheim, J. W., Shepson, P. B., Finlayson-Pitts, B. J., and Spicer, C. W.: The role of Br_2 and BrCl in surface ozone destruction at polar sunrise, Science, 291, 471–474, 2001.

Frinak, E. K. and Abbatt, J. P. D.: Br_2 production from the heterogeneous reaction of gas-phase OH with aqueous salt solutions: Impacts of acidity, halide concentration, and organic surfactants, J. Phys. Chem. A, 110, 10 456–64, 2006.

Gallard, H. and von Gunten, U.: Chlorination of natural organic matter: kinetics of chlorination and of THM formation, Water Res, 36, 65–74, 2002.

Gao, S., Keywood, M., Ng, N. L., Surratt, J. D., Varutbangkul, V., Bahreini, R., Flagan, R. C., and Seinfeld, J. H.: Low-molecular-weight and oligomeric components in secondary organic aerosol from the ozonolysis of cycloalkenes and α-pinene, J. Phys. Chem. A, 108, 10 147–10 164, 2004a.

Gao, S., Ng, N. L., Keywood, M., Varutbangkul, V., Bahreini, R., Nenes, A., He, J., Yoo, K. Y., Beauchamp, J. L., Hodyss, R. P., Flagan, R. C., and Seinfeld, J. H.: Particle phase acidity and oligomer formation in secondary organic aerosol, Environ. Sci. Technol., 38, 6582–6589, 2004b.

Gaspar, A., Kunenkov, E., Lock, R., Desor, M., Perminova, I., and Schmitt-Kopplin, P.: Combined utilization of ion mobility- and ultra high resolution-MS to identify multiply charged constituents in natural organic matter, Rapid Com. Mass Spec., 23, 683–688, 2009.

Gelencser, A., Meszaros, T., Blazso, M., Kiss, G., Krivacsy, Z., Molnar, A., and Meszaros, E.: Structural characterisation of organic matter in fine tropospheric aerosol by pyrolysis-gas chromatography-mass spectrometry, J. Atmos. Chem., 37, 173–183, 2000.

Gelencser, A., Hoffer, A., Kiss, G., Tombácz, E., Kurdi, R., and Bencze, L.: In-situ formation of light-absorbing organic matter in cloud water, J. Atmos. Chem., 45, 25–33, 2003.

George, I. J. and Abbatt, J. P. D.: Heterogeneous oxidation of atmospheric aerosol particles by gas-phase radicals, Nat. Chem., 2, 713–722, 2010.

References

Ghauch, A., Deveau, P.-A., Jacob, V., and Baussand, P.: Use of FTIR spectroscopy coupled with ATR for the determination of atmospheric compounds, Talanta, 68, 1294–1302, 2006.

Gong, S., Barrie, L., and Lazare, M.: Canadian Aerosol Module (CAM): A size-segregated simulation of atmospheric aerosol processes for climate and air quality models 2. Global sea-salt aerosol and its budgets, J. Geophys. Res., 107(D24), 4779–4792, 2002.

Graber, E. R. and Rudich, Y.: Atmospheric HULIS: how humic-like are they? A comprehensive and critical review, Atmos. Chem. Phys., 6, 729–753, 2006.

Griffiths, P. R. and de Haseth, J. A.: Fourier transform infrared spectroscopy, John Wiley & Sons, Inc., Hoboken, New Jersey, 2007.

Griffiths, P. R. and Homes, C. C.: Instrumentation for far-infrared spectroscopy. Handbook of vibrational spectroscopy., John Wiley & Sons Ltd, 2001.

Grothe, H.: II-8 Laboratory simulations, in: Spectroscopy of the atmospheres, edited by Escribano, R. and Tanarro, I., pp. 125–138, CSIC Publishing, Madrid, 2010.

Hallquist, M., Wenger, J. C., Baltensperger, U., Rudich, Y., Simpson, D., Claeys, M., Dommen, J., Donahue, N. M., George, C., Goldstein, A. H., Hammilton, J. F., Herrmann, H., Hoffmann, T., Iinuma, Y., Jang, M., Jenkin, M. E., Jimenez, J. L., Kiendler-Scharr, A., Maenhaut, W., McFiggans, G., Mentel, T. F., Monod, A., Prevot, A. S. H., Seinfeld, J. H., Surratt, J. D., Szmigielski, R., and Wildt, J.: The formation, properties and impact of secondary organic aerosol: current and emerging issues, Atmos. Chem. Phys., 9, 5155–5236, 2009.

Hatch, C. D., Gierlus, K. M., Zahardis, J., Schuttlefield, J., and Grassian, V. H.: Water uptake of humic and fulvic acid: measurments and modelling using single parameter Köhler theory, Environ. Chem., 6, 380–388, 2009.

Havers, N., Burba, P., Lambert, J., and Klockow, D.: Spectroscopic characterization of humic-like substances in airborne particulate matter, J. Atmos. Chem., 29, 45–54, 1998.

Hays, M. D., Fine, P. M., Geron, C. D., Kleeman, M. J., and Gullett, B. K.: Open burning of agricultural biomass: Physical and chemical properties of particle-phase emissions, Atmos. Environ., 39, 6747–6764, 2005.

Heaton, K. J., Dreyfus, M. A., Wang, S., and Johnston, M. V.: Oligomers in the early stage of biogenic secondary organic aerosol formation and growth, Environ. Sci. Technol., 41, 6129–36, 2007.

References

Henze, D. K., Seinfeld, J. H., Ng, N. L., Kroll, J. H., Fu, T. M., Jacob, D. J., and Heald, C. L.: Global modeling of secondary organic aerosol formation from aromatic hydrocarbons: high- vs. low-yield pathways, Atmos. Chem. Phys., 8, 2405–2420, 2008.

Herriott, D., Kogelnik, H., and Kompfner, R.: Off-axis paths in spherical mirror interferometers, Appl. Opt., 3, 523–526, 1964.

Hertkorn, N., Meringer, M., Gugisch, R., Ruecker, C., Frommberger, M., Perdue, E., Witt, M., and Schmitt-Kopplin, P.: High-precision frequency measurements: indispensable tools at the core of molecular-level analysis of complex systems, Anal. Bioanal. Chem., 389, 1311–1327, 2007.

Hertkorn, N., Frommberger, M., Schmitt-Kopplin, P., Witt, M., Koch, B., and Perdue, E.: Natural organic matter and the event horizon of mass spectrometry, Anal. Chem., 80, 8908–8919, 2008.

Hesse, M., Meier, H., and Zeeh, B.: Spektroskopische Methoden in der oganischen Chemie, Georg Thieme Verlag, Stuttgart, 1991.

Hoffer, A., Kiss, G., Blazso, M., and Gelencser, A.: Chemical characterization of humic-like substances (HULIS) formed from a lignin-type precursor in model cloud water, Geophys. Res. Lett., 31, L06115, 2004.

Hoffer, A., Gelencser, A., Guyon, P., Kiss, G., Schmid, O., Frank, G. P., Artaxo, P., and Andreae, M. O.: Optical properties of humic-like substances (HULIS) in biomass-burning aerosols, Atmos. Chem. Phys., 6, 3563–3570, 2006.

Holzinger, R., Kasper-Giebl, A., Staudinger, M., Schauer, G., and Röckmann, T.: Analysis of the chemical composition of organic aerosol at the Mt. Sonnblick observatory using a novel high mass resoltuion thermal-desorption proton-transfer-reaction-mass-spectrometer (hr-TD-PTR-MS), Atmos. Chem. Phys., 10, 10111–10128, 2010.

Huang, W.-J., Chen, L.-Y., and Peng, H.-S.: Effect of NOM characteristics on brominated organics formation by ozonation, Environ. Int., 29, 1049–1055, 2004.

Huber, S. G., Wunderlich, S., Schöler, H. F., and Williams, J.: Natural abiotic formation of furans in soil, Environ. Sci. Technol., 44, 5799–5804, 2010.

Iinuma, Y., Böge, O., Gnauk, T., and Herrmann, H.: Aerosol-chamber study of the α -pinene/O_3 reaction: influence of particle acidity on aerosol yields and products, Atmos. Environ., 38, 761–773, 2004.

Iinuma, Y., Böge, O., Kahnt, A., and Herrmann, H.: Laboratory chamber studies on the formation of organosulfates from reactive uptake of monoterpene oxides, Phys. Chem. Chem. Phys., 11, 7985–7997, 2009.

Ishikawa, T., Sato, T., Ose, Y., and Nagase, H.: Reaction of chlorine and bromine with humic substance, Sci. Total Environ., 54, 185–94, 1986.

Jang, M. and Kamens, R. M.: Characterization of secondary aerosol from the photooxidation of toluene in the presence of NO_x and 1-propene, Environ Sci Technol, 35, 3626–3639, 2001.

Jimenez, J. L., Canagaratna, M. R., Donahue, N. M., Prevot, A. S. H., Zhang, Q., Kroll, J. H., DeCarlo, P. F., Allan, J. D., Coe, H., Ng, N. L., Aiken, A. C., Docherty, K. S., Ulbrich, I. M., Grieshop, A. P., Robinson, A. L., Duplissy, J., Smith, J. D., Wilson, K. R., Lanz, V. A., Hueglin, C., Sun, Y. L., Tian, J., Laaksonen, A., Raatikainen, T., Rautiainen, J., Vaattovaara, P., Ehn, M., Kulmala, M., Tomlinson, J. M., Collins, D. R., Cubison, M. J., Dunlea, J., Huffman, J. A., Onasch, T. B., Alfarra, M. R., Williams, P. I., Bower, K., Kondo, Y., Schneider, J., Drewnick, F., Borrmann, S., Weimer, S., Demerjian, K., Salcedo, D., Cottrell, L., Griffin, R., Takami, A., Miyoshi, T., Hatakeyama, S., Shimono, A., Sun, J. Y., Zhang, Y. M., Dzepina, K., Kimmel, J. R., Sueper, D., Jayne, J. T., Herndon, S. C., Trimborn, A. M., Williams, L. R., Wood, E. C., Middlebrook, A. M., Kolb, C. E., Baltensperger, U., and Worsnop, D. R.: Evolution of organic aerosols in the atmosphere, Science, 326, 1525–1529, 2009.

Johnson, D., Jenkin, M. E., Wirtz, K., and Martin-Reviejo, M.: Simulating the formation of secondary organic aerosol from the photooxidation of aromatic hydrocarbons, Environ. Chem., 2, 35–48, 2005.

Johnson, S. A., Kumar, R., and Cunningham, P. T.: Airborne detection of acidic sulfate aerosol using an ATR impactor, Aerosol Sci. Technol., 2, 401–5, 1983.

Jonsson, A. M., Hallquist, M., and Saathoff, H.: Volatility of secondary organic aerosols from the ozone initiated oxidation of α-pinene and limonene, J. Aerosol Sci., 38, 843–852, 2007.

Jonsson, A. M., Hallquist, M., and Ljungström, E.: The effect of temperature and water on secondary organic aerosol formation from ozonolysis of limonene, Δ^3-carene and α-pinene, Atmos. Chem. Phys., 8, 6541–6549, 2008.

Juranyi, Z., Gysel, M., Duplissy, J., Weingartner, E., Tritscher, T., Dommen, J., Henning, S., Ziese, M., Kiselev, A., Stratmann, F., George, I., and Baltensperger, U.: Influence of gas-to-particle partitioning on the hygroscopic and droplet activation behaviour of α-pinene secondary organic aerosol, Phys. Chem. Chem. Phys., 11, 8091–8097, 2009.

References

Kalberer, M., Paulsen, D., Sax, M., Steinbacher, M., Dommen, J., Prevot, A. S. H., Fisseha, R., Weingartner, E., Frankevich, V., Zenobi, R., and Baltensperger, U.: Identification of polymers as major components of atmospheric organic aerosols, Science, 303, 1659–1662, 2004.

Kanakidou, M., Seinfeld, J. H., Pandis, S. N., Barnes, I., Dentener, F. J., Facchini, M. C., Van Dingenen, R., Ervens, B., Nenes, A., Nielsen, C. J., Swietlicki, E., Putaud, J. P., Balkanski, Y., Fuzzi, S., Horth, J., Moortgat, G. K., Winterhalter, R., Myhre, C. E. L., Tsigaridis, K., Vignati, E., Stephanou, E. G., and Wilson, J.: Organic aerosol and global climate modelling: a review, Atmos. Chem. Phys., 5, 1053–1123, 2005.

Khovratovich, N. N., Novikova, T. M., Khmel'nitskii, A. I., Cherenkevich, S. N., and Loban, V. A.: IR spectra of preparations of ozonized pyrocatechin, J. Appl. Spectrosc., 65, 201–205, 1998.

Kiendler-Scharr, A., Wildt, J., Maso, M. D., Hohaus, T., Kleist, E., Mentel, T. F., Tillmann, R., Uerlings, R., Schurr, U., and Wahner, A.: New particle formation in forests inhibited by isoprene emissions, Nature, 461, 381–384, 2009.

Knutson, E. O. and Whitby, K. T.: Aerosol classification by electric mobility: Apparatus, theory, and applications, J. Aerosol Sci., 6, 443–451, 1975.

Kodas, T. T., Pratsinis, S. E., and Friedlander, S. K.: Aerosol formation and growth in a laminar core reactor, J. Colloid Interface Sci., 111, 102–11, 1986.

Kopetzky, R. and Palm, W.-U.: Formation of halogenated compounds from humic acids in saline solutions, Tech. rep., Institut für Ökologie und Umweltchemie - Universität Lüneburg, 2006.

Kortüm, G. and Oelkrug, D.: Reflexionsspektren fester Stoffe, Naturwissenschaften, 53, 600–609, 1966.

Kortüm, G., Braun, W., and Herzog, G.: Prinzip und Messmethodik der diffusen Reflexionsspektroskopie, Angew. Chem., 75, 653–696, 1963.

Kroll, J. H. and Seinfeld, J. H.: Chemistry of secondary organic aerosol: Formation and evolution of low-volatility organics in the atmosphere, Atmos. Environ., 42, 3593–3624, 2008.

Kroll, J. H., Donahue, N. M., Jimenez, J. L., Kessler, S. H., Canagaratna, M. R., Wilson, K. R., Altieri, K., Mazzoleni, L., Wozniak, A., Bluhm, H., Mysak, E., Smith, J. D., Kolb, C. E., and Worsnop, D. R.: Carbon oxidation state as a metric for describing the chemistry of atmospheric organic aerosols, nature chemistry, 3, 133–139, 2011.

Kubala, D., Drage, E., Al-Faydhi, A., Kocisek, J., Pappa, P., Matejcik, V., Mach, P., Urban, J., Limao-Vieira, P., Hoffmann, S., Matejcik, S., and Mason, N.: Elektron impact ionisation and UV absorption study of α- and β-pinene, Int. J. Mass Spectrom., 280, 169–173, 2009.

Kubelka, P. and Munk, F.: Ein Beitrag zur Optik der Farbanstriche, Z. tech. Phys., 12, 593–601, 1931.

Kumagai, K., Iijima, A., Shimoda, M., Saitoh, Y., Kozawa, K., Hagino, H., and Sakamoto, K.: Determination of dicarboxylic acids and levoglucosan in fine particles in the Kanto Plain, Japan, for source apportionment of organic aerosols, Aerosol and Air Quality Research, 10, 282–291, 2010.

Kundu, S., Kawamura, K., Andreae, T. W., Hoffer, A., and Andreae, M. O.: Molecular distributions of dicarboxylic acids, ketocarboxylic acids and α-dicarbonyls in biomass burning aerosols: implications for photochemical production and degradation in smoke layers, Atmos. Chem. Phys., 10, 2209–2225, 2010.

Lary, D., Shallcross, D., and Toumi, R.: Carbonaceous aerosols and their potential role in atmospheric chemistry, Journal of Geophysical Research, 104, 15 929–15 940, 1999.

Last, D. J., Najera, J. J., Percival, C. J., and Horn, A. B.: A comparison of infrared spectroscopic methods for the study of heterogeneous reactions occurring on atmospheric aerosol proxies, Phys. Chem. Chem. Phys., 11, 8214–8225, 2009.

Lay, T. H. and Bozzelli, W.: Atmospheric photocheical oxidation of benzene: Benzene + OH and the benzene-OH adduct (Hydroxyl-2,4-cyclohexadienyl) + O_2, J. Phys. Chem., 100, 6543–6554, 1996.

Lim, Y. B. and Ziemann, P. J.: Kinetics of the heterogeneous conversion of 1,4-hydroxycarbonyls to cyclic hemiacetals and dihydrofurans on organic aerosol particles, Phys. Chem. Chem. Phys., 11, 8029–8039, 2009.

Limbeck, A. and Puxbaum, H.: Organic acids in continental background aerosols, Atmos. Environ., 33, 1847–1852, 1999.

Limbeck, A., Kraxner, Y., and Puxbaum, H.: Gas to particle distribution of low molecular weight dicarboxylic acids at two different sites in central Europe (Austria), J. Aerosol Sci., 36, 991–1005, 2005.

Mainelis, G., Adhikari, A., Willeke, K., Lee, S.-A., Reponen, T., and Grinshpun, S. A.: Collection of airborne microorganisms by a new electrostatic precipitator, J. Aerosol Sci., 33, 1417–1432, 2002.

References

Maksymiuk, C. S., Gayahtri, C., Gil, R. R., and Donahue, N. M.: Secondary organic aerosol formation from multiphase oxidation of limonene by ozone: mechanistic constraints via two-dimensional heteronuclear NMR spectroscopy, Phys. Chem. Chem. Phys., 11, 7810–7818, 2009.

McDow, S. R., Sun, Q.-R., Vartiainen, M., Hong, Y.-S., Yao, Y.-l., Fister, T., Yao, R.-Q., and Kamens, R. M.: Effect of composition and state of organic components on polycyclic aromatic hydrocarbon decay in atmospheric aerosols, Environ. Sci. Technol., 28, 2147–53, 1994.

Moise, T. and Rudich, Y.: Uptake of Cl and Br by organic surfaces - A perspective on organic aerosols processing by tropospheric oxidants, Geophys. Res. Lett., 28, 4083–4086, 2001.

Molina, L. and Rowland, F. S.: Stratospheric sink for chlorofluoromethanes: Chlorine atom-catalyzed destruction of ozone, Nature, 249, 810–812, 1974.

Moosmüller, H., Chakrabarty, R., Ehlers, K., and Arnott, W.: Absorption Ångström coefficient, brown carbon, and aerosols: basic concepts, bulk matter, and spherical particles, Atmos. Chem. Phys., 11, 1217–1225, 2011.

Morrow, P. E. and Mercer, T. T.: A point-to-plane electrostatic precipitator for particle size sampling, Am. Ind. Hyg. Assoc. J., 25, 8–14, 1964.

Mosher, B., Winkler, P., and Jaffrezo, J. L.: Seasonal aerosol chemistry at Dye 3, Greenland, Atmos. Environ., 27A, 2761–2772, 1993.

Muckenhuber, H. and Grothe, H.: The reaction between soot and NO_2 - investigation on functional groups using TPD-MS, Topics in Catalysis, 30/31, 287–291, 2004.

Muckenhuber, H. and Grothe, H.: The heterogeneous reaction between soot and NO_2 at elevated temperature, Carbon, 44, 546–559, 2006.

Muckenhuber, H. and Grothe, H.: A DRIFTS study of the heterogeneous reaction of NO_2 with carbonaceous materials at elevated temperature, Carbon, 45, 321–329, 2007.

Mukai, H. and Ambe, Y.: Characterization of a humic acid-like brown substance in airborne particulate matter and tentative identification of its origin, Atmos. Environ., 20, 813–819, 1986.

Myhre, C. E. L., D'Anna, B., Nicolaisen, F. M., and Nielsen, C. J.: Properties of aqueous methanesulfonic acid: complex index of refraction and surface tension, Appl. Opt., 43, 2500–2509, 2004.

Najera, J. J., Fochesatto, J. G., Last, D. J., Percival, C. J., and Horn, A. B.: Infrared spectroscopic methods for the study of aerosol particles using White cell optics: Development and characterization of a new aerosol flow tube, Rev. Sci. Instrum., 79, 124 102/1–124 102/12, 2008.

Najera, J. J., Percival, C. J., and Horn, A. B.: Infrared spectroscopic studies of the heterogeneous reaction of ozone with dry maleic and fumaric acid aerosol particles, Phys. Chem. Chem. Phys., 11, 9093–9103, 2009.

Ng, N. L., Kroll, J. H., Chan, A. W. H., Chhabra, P. S., Flagan, R. C., and Seinfeld, J. H.: Secondary organic aerosol formation from m-xylene, toluene, and benzene, Atmos. Chem. Phys., 7, 3909–3922, 2007.

Ng, N. L., Canagaratna, M. R., Zhang, Q., Jimenez, J. L., Tian, J., Ulbrich, I. M., Kroll, J. H., Docherty, K. S., Chhabra, P. S., Bahreini, R., Murphy, S. M., Seinfeld, J. H., Hildebrandt, L., Donahue, N. M., DeCarlo, P. F., Lanz, V. A., Prévôt, A. S. H., Dinar, E., Rudich, Y., and Worsnop, D. R.: Organic aerosol components observed in northern hemispheric datasets from aerosol mass spectrometry, Atmos. Chem. Phys., 10, 4625–4641, 2010.

Nießner, R.: Chemische Charakterisierung von Aerosolen: "online" und "in situ", Angew. Chem., 103, 542–552, 1991.

Nieto-Gligorovski, L., Net, S., Gligorovski, S., Zetzsch, C., Jammoul, A., D'Anna, B., and George, C.: Interactions of ozone with organic surface films in the presence of simulated sunlight: impact on wettability of aerosols, Phys. Chem. Chem. Phys., 10, 2964–2971, 2008.

Nieto-Gligorovski, L., Net, S., Gligorovski, S., Wortham, H., Grothe, H., and Zetzsch, C.: Spectroscopic study of organic coatings on fine particles, exposed to ozone and simulated sunlight, Atmos. Environ., 44, 5451–5459, 2010.

Nolting, F. and Zetzsch, C.: The ozone forming potential of the biogenic α-pinene in comparison with anthropogenic hydrocarbons, in: Ozone in the Atmosphere. Proc. Quadrennial Ozone Symposium 1988, edited by Bojkov, R. and Fabian, P., Deepak, Hampton, VA, 1989.

Nolting, F., Behnke, W., and Zetzsch, C.: A smog chamber for studies of the reactions of terpenes and alkanes with ozone and hydroxyl, J. Atmos. Chem., 6, 47–59, 1988.

Offenberg, J. H., Lewis, C. W., Lewandowski, M., Jaoui, M., Kleindienst, T. E., and Edney, E. O.: Contributions of toluene and α-pinene to SOA formed in an irradiated toluene/α-pinene/NO_x/ air mixture: Comparison of results using ^{14}C content and SOA organic tracer methods, Environ. Sci. Technol., 41, 3972–3976, 2007.

References

Ofner, J.: Mechanismus heterogener Reaktionen an der Russoberfläche, Institute of Materials Chemistry, Vienna University of Technology, Vienna, 2006.

Ofner, J. and Grothe, H.: A mechanistic study of the cooperative effect of NO_2 and O_2 on the soot surface, Asian Chem. Let., 11, 57–61, 2007.

Ofner, J., Krüger, H.-U., Zetzsch, C., and Grothe, H.: Direct deposition of aerosol particles on an ATR crystal for FTIR spectroscopy using an electrostatic precipitator, Aerosol Sci. Technol., 43, 1–5, 2009.

Ofner, J., Krüger, H.-U., and Zetzsch, C.: Time resolved infrared spectroscopy of formation and processing of secondary organic aerosols, Z. Phys. Chem., 224, 1171–1183, 2010a.

Ofner, J., Krüger, H.-U., and Zetzsch, C.: Circular multireflection cell for optical spectroscopy, Appl. Opt., 49, 5001–5004, 2010b.

Ofner, J., Krüger, H.-U., Grothe, H., Schmitt-Kopplin, P., Whitmore, K., and Zetzsch, C.: Physico-chemical characterization of SOA derived from catechol and guaiacol - a model substance for the aromatic fraction of atmospheric HULIS, Atmos. Chem. Phys., 11, 1–15, 2011.

Olariu, R. I., Barnes, I., Becker, K. H., and Klotz, B.: Rate coefficients for the gas-phase reaction of OH radicals with selected dihydroxybenzenes and benzoquinones, Int. J. Chem. Kinet., 32, 696–702, 2000.

Olariu, R. I., Tomas, A., Barnes, I., Bejan, I., Becker, K. H., and Wirtz, K.: Atmospheric ozone degradation reaction of 1,2-dihydroxybenzene: Aerosol formation study, The European Photoreactor EUPHORE, 4th Report 2001, 2004.

Pan, X., Underwood, J. S., Xing, J.-H., Mang, S. A., and Nizkorodov, S. A.: Photodegradation of secondary organic aerosol generated from limonene oxidation by ozone studied with chemical ionization mass spectrometry, Atmos. Chem. Phys., 9, 3851–3865, 2009.

Pandis, S. N., Paulson, S., Flagan, R. C., and Seinfeld, J. H.: Aerosol formation in the photooxidation of isoprene and β-pinene, Atmos. Environ., 25A, 997–1008, 1991.

Paulot, F., Crounse, J. D., Kjaergaard, H. G., Kuerten, A., St. Clair, J. M., Seinfeld, J. H., and Wennberg, P. O.: Unexpected epoxide formation in the gas-phase photooxidation of isoprene, Science, 325, 730–733, 2009.

Perner, D. and Platt, U.: Detection of nitrous acid in the atmosphere by differential optical absorption, Geophys. Res. Lett., 6, 917–920, 1979.

Platt, U.: Reactive halogen species in the mid-latitude troposphere - recent discoveries, Water, Air, Soil Pollut., 123, 229–244, 2000.

Platt, U. and Hönninger, G.: The role of halogen species in the troposphere, Chemosphere, 52, 325–338, 2003.

Polidori, A., Turpin, B. J., Davidson, C. I., Rodenburg, L. A., and Maimone, F.: Organic PM2.5: Fractionation by polarity, FTIR spectroscopy, and OM/OC ratio for the Pittsburgh aerosol, Aerosol Sci. Technol., 42, 233–246, 2008.

Pöschl, U.: Atmospheric aerosols: Composition, transformation, climate and health effects, Angew. Chem., Int. Ed., 44, 7520–7540, 2005.

Pöschl, U., Martin, S. T., Sinha, B., Chen, Q., Gunthe, S. S., Huffman, J. A., Borrmann, S., Farmer, D. K., Garland, R. M., Helas, G., Jimenez, J. L., King, S. M., Manzi, A., Mikhailov, E., Pauliquevis, T., Petters, M. D., Prenni, A. J., Roldin, P., Rose, D., Schneider, J., Su, H., Zorn, S. R., Artaxo, P., and Andreae, M. O.: Rainforest aerosols as biogenic nuclei of clouds and precipitation in the Amazon, Science, 329, 1513–1516, 2010.

Rahn, K. A., Borys, R. D., Butler, E. L., and Duce, R. A.: Gaseous and particulate halogens in the New York city atmosphere, Ann. NY Acad. Sci., pp. 143–151, 1979.

Ramaswamy, V., Boucher, O., Haigh, J., Hauglustaine, D., Haywood, J., Myhre, G., Nakajima, T., Shi, G., and Solomon, S.: Radiative Forcing of Climate Change, in: Climate Change 2001: The Scientific Basis. Contribution of Working Group I to the Third Assessment Report of the Intergovernmental Panel on Climate Change, edited by Houghton, J., Ding, Y., Griggs, D., Noguer, M., van der Linden, P., Dai, X., Maskell, K., and Johnson, C., Cambridge University Press, United Kingdom and New York, 2001.

Reinhardt, A., Emmenegger, C., Gerrits, B., Panse, C., Dommen, J., Baltensperger, U., Zenobi, R., and Kalberer, M.: Ultra-high mass resolution and accurate mass measurements as new tools to characterize oligomers in secondary organic aerosol, Anal. Chem., 79, 4074–4082, 2007.

Reist, P. C.: Aerosol Science and Technology, McGraw-Hill, Inc., New York, 1993.

Remorov, R., Zasetsky, A., and Sloan, J.: Low pressure aerosol flow reactor, Aerosol Sci. Technol., 39, 1038–1047, 2005.

Robert, C.: Simple, stable, and compact multiple-reflection optical cell for very long optical paths, Appl. Opt., 46, 5408–18, 2007.

References

Robinson, A. L., Donahue, N. M., Shrivastava, M. K., Weitkamp, A., Sage, A. M., Grieshop, A. P., Lane, T. E., Pierce, J. R., and Pandis, S. N.: Rethinking organic aerosols: Semivolatile emissions and photochemical aging, Science, 315, 1259–1262, 2007.

Rontu, N. and Vaida, V.: Vibrational spectroscopy of perfluorocarboxylic acids from the infrared to the visible regions, J. Phys. Chem. B, 112, 276–282, 2008.

Rossi, M. J.: Heterogeneous reactions on salts, Chem Rev, 103, 4823–82, 2003.

Rowland, F. S.: Stratospheric ozone depletion, Phil. Trans. R. Soc. B, 361, 769–790, 2006.

Rowland, F. S. and Molina, L.: Chlorofluoromethanes in the environment, Rev. Geophys. Space Phys., 13, 1–35, 1975.

Rudich, Y.: Laboratory perspectives on the chemical transformations of organic matter in atmospheric particles, Chem. Rev., 103, 5097–124, 2003.

Salma, I. and Lang, G. G.: How many carboxyl groups does an average molecule of humic-like substances contain?, Atmos. Chem. Phys., 8, 5997–6002, 2008.

Salma, I., Meszaros, T., Maenhaut, W., Vass, E., and Majer, Z.: Chirality and the origin of atmospheric humic-like substances, Atmos. Chem. Phys., 10, 1315–1327, 2010.

Samburova, V., Didenko, T., Kunenkov, E., Emmenegger, C., Zenobi, R., and Kalberer, M.: Functional group analysis of high-molecular weight compounds in the water-soluble fraction of organic aerosols, Atmos. Environ., 41, 4703–4710, 2007.

Sax, M., Zenobi, R., Baltensperger, U., and Kalberer, M.: Time resolved infrared spectroscopic analysis of aerosol formed by photo-oxidation of 1,3,5-trimethylbenzene and α-pinene, Aerosol Sci. Technol., 39, 822–830, 2005.

Schnelle-Kreis, J., Sklorz, M., Herrmann, H., Zimmermann, R., and Schnelle-Kreis, J.: Sources, occurrence, composition. Atmospheric aerosols, Chem. Unserer Zeit, 41, 220–225,227–230, 2007.

Seinfeld, J. H. and Pandis, S. N.: Atmospheric chemistry and physics: from air pollution to climate change, John Wiley & Sons, Inc., 2nd. ed. edn., 2006.

Seinfeld, J. H., Kleindienst, T. E., Edney, E. O., and Cohen, J. B.: Aerosol growth in a steady-state, continuous flow chamber: Application to studies of secondary aerosol formation, Aerosol Sci. Technol., 37, 728–734, 2003.

References

Shapiro, E. L., Szprengiel, J., Sareen, N., Jen, C. N., Giordano, M. R., and McNeill, V. F.: Light-absorbing secondary organic material formed by glyoxal in aqueous aerosol mimics, Atmos. Chem. Phys., 9, 2289–2300, 2009.

Shevchenko, L. L.: Infrared spectra of salts and complexes of carboxylic acids and some of their derivatives, Russ. Chem. Rev., 32, 201–207, 1963.

Shimanouchi, T.: Tables of molecular vibrational frequencies: Consolidated Volume I, Tech. rep., United States department of commerce - National bureau of standards, 1972.

Siebert, H.: Anwendungen der Schwingungsspektroskopie in der anorganischen Chemie, Springer Verlag Berlin-Heidelberg-New York, 1966.

Siekmann, F.: Freisetzung von photolabilen und reaktiven Halogenverbindungen aus salzhaltigen Aerosolen unter simulierten und tropospärischen Reinluftbedingungen in einer Aerosol-Smogkammer, Ph.D. thesis, University of Bayreuth, 2008.

Siese, M., Becker, K. H., Brockmann, K. J., Geiger, H., Hofzumahaus, A., Holland, F., Mihelcic, D., and Wirtz, K.: Direct measurement of OH radicals from ozonolysis of selected alkenes: A EUPHORE simulation chamber study, Environ. Sci. Technol., 35, 4660–4667, 2001.

Simpson, W. R., von Glasow, R., Riedel, K., Anderson, P., Ariya, P., Bottenheim, J., Burrows, J., Carpenter, L. J., Friess, U., Goodsite, M. E., Heard, D., Hutterli, M., Jacobi, H. W., Kaleschke, L., Neff, B., Plane, J., Platt, U., Richter, A., Roscoe, H., Sander, R., Shepson, P., Sodeau, J., Steffen, A., Wagner, T., and Wolff, E.: Halogens and their role in polar boundary-layer ozone depletion, Atmos. Chem. Phys., 7, 4375–4418, 2007.

Smoydzin, L. and von Glasow, R.: Do organic surface films on sea salt aerosols influence atmospheric chemistry? - a model study, Atmos. Chem. Phys., 7, 5555–5567, 2007.

Socrates, G.: Infrared characteristic group frequencies, John Wiley & Sons Chichester-New York-Brisbane-Toronto, 1980.

Solomon, S.: Stratospheric ozone depletion: A review of concepts and history, Rev. Geophys., 37, 275–316, 1999.

Sonntag, D.: Important new values of the physical constants of 1986, vapor pressure formulations based on the ITS-90 and psychrometer formulae, Z. Meteorol., 70, 340–344, 1990.

Sörgel, M.: Experimentelle Untersuchung zur Bildung halogenorganischer Verbindungen aus Huminsäuren in Abhängigkeit vom pH-Wert, Fachbereich Naturwissenschaftliche Technik, Hochschule für Angewandte Wissenschaften, Hamburg, 2007.

References

Steinbrecher, R. and Koppmann, R.: Important biogenic hydrocarbons. Biosphere and atmosphere, Chem. Unserer Zeit, 41, 286–292, 2007.

Stevenson, F.: Humus chemistry: genesis, composition and reactions, Wiley, New York, 2nd ed. edn., 1994.

Stevenson, F. J. and Goh, K. M.: Infrared spectra of humic acids and related substances, Geochim. Cosmochim. Ac., 35, 471–483, 1971.

Stone, E. A., Hedman, C. J., Sheesley, R. J., Shafer, M. M., and Schauer, J. J.: Investigating the chemical nature of humic-like substances (HULIS) in North American atmospheric aerosols by liquid chromatography tandem mass spectrometry, Atmos. Environ., 43, 4205–4213, 2009.

Sun, Y. L., Zhang, Q., Anastasio, C., and Sun, J.: Insights into secondary organic aerosol formed via aqueous-phase reactions of phenolic compounds based on high resolution mass spectrometry, Atmos. Chem. Phys., 10, 4809–4822, 2010.

Tas, E., Peleg, M., Pedersen, D. U., Matveev, V., Pour Biazar, A., and Luria, M.: Measurement-based modeling of bromine chemistry in the boundary layer: 1. Bromine chemistry at the Dead Sea, Atmos. Chem. Phys., 6, 5589–5604, 2006.

Thoma, M. L., Kaschow, R., and Hindelang, F.: A multiple-reflection cell suited for absorption measurements in shock tubes, Shock Waves, 4, 51–53, 1994.

Tomas, A., Olariu, R. I., Barnes, I., and Becker, K. H.: Kinetics of the reaction of O_3 with selected benzenediols, Int. J. Chem. Kinet., 35, 223–230, 2003.

Toyota, K., Kanaya, Y., Takahashi, M., and Akimoto, H.: A box model study on photochemical interactions between VOCs and reactive halogen species in the marine boundary layer, Atmos. Chem. Phys., 4, 1961–1987, 2004.

Tretyakova, N. Y., Lebedev, A. T., and Petrosyan, V. S.: Degradative pathways for aqueous chlorination of orcinol, Environ. Sci. Technol., 28, 606–13, 1994.

Uyguner, C. S., Hellriegel, C., Otto, W., and Larive, C. K.: Characterization of humic substances: Implications for trihalomethane formation, Anal. Bioanal. Chem., 378, 1579–1586, 2004.

Vesna, O., Sax, M., Kalberer, M., Gaschen, A., and Ammann, M.: Product study of oleic acid ozonolysis as function of humidity, Atmos. Environ., 43, 3662–3669, 2009.

Virtanen, A., Joutsensaari, J., Koop, T., Kannosto, J., Yli-Pirilä, P., Leskinen, J., Mäkelä, J. M., Holopainen, J. K., Pöschl, U., Kulmala, M., Worsnop, D. R., and Laaksonen, A.: An amorphous solid state of biogenic secondary organic aerosol particles, nature, 467, 824–827, 2010.

Von Glasow, R. and Crutzen, P. J.: Tropospheric halogen chemistry, Treatise Geochem., 4, 21–64, 2004.

Wagner, T., Ibrahim, O., Sinreich, R., Friess, U., von Glasow, R., and Platt, U.: Enhanced tropospheric BrO over antarctic sea ice in mid winter observed by MAX-DOAS on board the research vessel Polarstern, Atmos. Chem. Phys., 7, 3129–3142, 2007.

White, J. U.: Long optical path of large aperture, J. O. S. A., 32, 285–288, 1942.

Wiedensohler, A.: An approximation of the bipolar charge distribution for particles in the submicron size range, J. Aerosol Sci., 19, 387–389, 1988.

Wilson, H. W.: The infrared and Raman spectra of α- and β-pinenes, Appl. Spectrosc., 30, 209–12, 1976.

Xu, D., Dan, M., Song, Y., Chai, Z., and Zhuang, G.: Concentration characteristics of extractable organohalogens in PM2.5 and PM10 in Beijing, China, Atmos. Environ., 39, 4119–4128, 2005.

Yu, Y., Ezell, M. J., Zelenyuk, A., Imre, D., Alexander, L., Ortega, J., D'Anna, B., Harmon, C. W., Johnson, S. N., and Finlayson-Pitts, B. J.: Photooxidation of α-pinene at high relative humidity in the presence of increasing concentrations of NO_x, Atmos. Environ., 42, 5044–5060, 2008.

Zellner, R., Behr, P., Seisel, S., Somnitz, H., and Treuel, L.: Chemistry and microphysics of atmospheric aerosol surfaces: Laboratory techniques and applications, Z. Phys. Chem., 223, 359–385, 2009.

Zetzsch, C., Pfahler, G., and Behnke, W.: Heterogeneous formation of chlorine atoms from NaCl in a photosmog system, J. Aerosol Sci., 19, 1203–1206, 1988.

Zhang, Y., Hu, Y., Ding, F., and Zhao, L.: FTIR-ATR chamber for observation of efflorescence and deliquescence processes of $NaClO_4$ aerosol particles on ZnSe substrate, Chin. Sci. Bull., 50, 2149–2152, 2005.

Zhao, F., Gong, Z., Hu, H., Tanaka, M., and Hayasaka, T.: Simultaneous determination of the aerosol complex index of refraction and size distribution from scattering measurements of polarized light, Appl. Opt., 36, 7992–8001, 1997.

List of Figures

1.1 Current understanding of the oxidation steps (gas phase: $ox.^g$; particle phase: $ox.^p$) starting from a SOA precursor (VOC) leading to low-volatile compounds in the gas phase (P_g), partitioning to the particle phase (P_p) and finally releasing simple gaseous molecules (G) like CO_2 or CO. Adapted from Kroll and Seinfeld (2008). 4

1.2 The terpene precursor isoprene (a) and the monoterpenes (-)-α-pinene (b) and (+)-α-pinene (c). The racemate of those two structural isomers, called α-pinene, is widely used as model compound in laboratory experiments for SOA research. 5

1.3 Some benzene-type precursors and selected literature on SOA formation: (a) benzene by Johnson et al. (2005), Baltensperger et al. (2005), Henze et al. (2008), and Ng et al. (2007); (b) toluene by Forstner et al. (1997), Henze et al. (2008), Ng et al. (2007), Seinfeld et al. (2003), and Jang and Kamens (2001); (c) m-xylene and (d) p-xylene by Ng et al. (2007), Forstner et al. (1997), Johnson et al. (2005), and Henze et al. (2008); (e) 1,3,5-trimethylbenzene by Johnson et al. (2005) and Baltensperger et al. (2005) 8

1.4 Some phenol-type precursors: (a) phenol; (b) catechol; (c) guaiacol 9

1.5 2-D plot of aerosol formation and aging; change of O/C ratio and saturation concentration C^* of the aerosol precursor (SOA-PC) by functionalization and oligomerization; formation of SV- and LV-OOA in the related O/C-C^* space. Adapted from Jimenez et al. (2009). 12

1.6 Different organic aerosol types (LV-OOA, SV-OOA, BBOA, WSOC, HOA (hydrocarbon-like organic aerosol)) and SOA precursors used in this study (catechol, guaiacol, α-pinene) in the average carbon oxidation state ($\overline{OS_C}$) and number of carbon atoms (n_C) space. Adapted from Kroll et al. (2011). 14

1.7 Simplified scheme of the halogen-release mechanism from sea-salt surfaces (according to Finlayson-Pitts (2010)) 16

1.8 Halogen release from a simulated salt pan in a Teflon smog chamber (with kind permission of N. Balzer and J. Buxmann) 17

List of Figures

1.9 Possible formation of halogenated SOA or BBOA (halo-SOA/BBOA), secondary halogen-organic aerosol (XOA), and halogenated organic gaseous species (VOX) by interaction of reactive halogen species released from salt lakes or sea-salt aerosol with SOA, BBOA, or their respective organic precursors (SOA-pc, BBOA-pc). 19

1.10 Examples of remote areas where organic-halogen interaction might take place. . 20

2.1 700 L aerosol smog chamber 23

2.2 Spectrum of the solar simulator Osram Metallogen HMI 4000 W with and without the water-cooled UV-C filter, compared to a calculated solar spectrum (40° N 11° E, maritime albedo, summer, no aerosol, 348 DU of ozone) 24

2.3 Setup of the aerosol flow reactor coupled to the FTIR spectrometer: a–flow reactor; b–movable inlet; c–impinger to vaporize the aerosol precursor; d & e–flow meter to control the gas flows; f–carrier gas inlet; g–reactive gas inlet; h–infrared gas cell; i–FTIR sample compartment; j–outlet. 26

2.4 Geometric concept and setup (photo by Christian Wißler, University of Bayreuth) of the circular multi-reflection cell 28

2.5 Path lengths achieved with the circular multi-reflection cell 28

2.6 Concept and picture of the experimental setup and calculation of the electric field of the electrostatic precipitator for coating ATR crystals–version 1 33

2.7 Concept and picture of the experimental setup, and calculation of the electric field inside the electrostatic precipitator for coating ATR crystals–version 2 . . . 35

2.8 Schematic figure of the vacuum system of the TPP-MS system at the Atmospheric Chemistry Research Laboratory of the University of Bayreuth 37

2.9 Internal nebulizer to add sea-salt aerosol to an existing aerosol in the aerosol smog chamber (developed by H.-U. Krüger) 40

3.1 Formation of SOA from α-pinene in the 700 L aerosol smog chamber 43

3.2 Time-resolved long-path infrared absorption spectra of SOA formation from α-pinene with ozone and without simulated sunlight 44

3.3 Formation of SOA from α-pinene studied inside an aerosol smog chamber (SC, temporal resolution 10 min) and an aerosol flow reactor (AFR, temporal resolution 1 s up to 10 s) 46

3.4 ATR-FTIR spectra of particulate matter of SOA from α-pinene at two different ambient conditions 47

3.5 TPP mass spectra of particulate matter of SOA from α-pinene 49

3.6 Diffuse reflectance spectra ($F(R)$) of SOA from α-pinene and absorption spectra of a saturated α-pinene vapor 50

List of Figures

3.7 Aerosol size distribution of SOA from catechol formed in the presence of simulated sunlight and 25 % relative humidity and formed in the dark at 0 % relative humidity 51

3.8 Comparison of the evolution of the aerosol mass concentration (C_{OA}) at different simulated environmental conditions and depending on the SOA precursors. . . . 52

3.9 FEG-SEM images of SOA from catechol, exhibiting two different morphologies . 53

3.10 Time-resolved long-path infrared absorption spectra of SOA formation from catechol with ozone and without simulated sunlight: The precursor (dash-dotted) rapidly decreases and the broad, largely unstructured bands of the organic aerosol (black) increase at the same time. 54

3.11 Long-path absorption infrared spectra of catechol and guaiacol compared to SOA formed after 30 minutes at different ambient conditions. 56

3.12 Infrared spectra of the formation of SOA from catechol studied in an aerosol smog chamber (SC, temporal resolution 10 min) and an aerosol flow reactor (AFR, temporal resolution 1 s) 59

3.13 Evolution of the carbonyl stretching region of SOA from catechol at varying catechol/ozone ratios: A–1:0.5, B–1:1, C–1:1.5, D–1:5 60

3.14 ATR-FTIR spectra of particulate matter of SOA from catechol and guaiacol at various ambient conditions 62

3.15 TPP-MS signals of the main masses for oxygen-containing functional groups observed as a function of temperature for the five different SOA 65

3.16 Diffuse-reflectance UV/VIS spectra of SOA from catechol and guaiacol, absorbance spectra of the precursors, and images of SOA samples 67

3.17 Ultra-high-resolution mass spectra (a) and Van Krevelen diagrams (b) of the different SOA samples 68

3.18 Examples of elemental compositions determined by ICR-FT/MS 70

4.1 Change of the mean particle diameter of organic aerosols by the reaction with gaseous halogens (L: SOA formed under simulated sunlight at 0 % relative humidity; LW: SOA formed under simulated sunlight at 25 % relative humidity), with the vertical lines marking the injection of the halogen after 60 minutes aerosol formation. 73

4.2 Long-path infrared spectroscopy of the heterogeneous reaction of the different organic aerosols with molecular halogens; L: SOA formed under simulated sunlight at 0 % relative humidity; LW: SOA formed under simulated sunlight at 25 % relative humidity) 76

List of Figures

4.3 ATR-FTIR spectra of the particulate phase of the aerosol after reaction with halogens; L: SOA formed at 0 % relative humidity and simulated sunlight; LW: SOA formed at 25 % relative humidity and simulated sunlight) 78

4.4 TPP-MS spectra of the masses 35, 37, 79 and 81, representing ^{35}Cl, ^{37}Cl, ^{79}Br, and ^{81}Br, and the calculated mass ratio (dotted) to determine thermal regions were the isotopic ratios can be found (L: SOA formed at 0 % relative humidity and simulated sunlight; LW: SOA formed at 25 % relative humidity and simulated sunlight) 80

4.5 TPP mass spectra of SOA from catechol (LW) between 75 and 85 amu: The masses of the halogen atoms (^{79}Br and ^{81}Br) correlate with those of HBr (^{80}M and ^{82}M). 81

4.6 Van Krevelen diagrams of halogenated SOA from catechol formed at 0 % relative humidity and simulated sunlight 82

4.7 Van Krevelen diagrams of halogenated SOA from guaiacol formed at 0 % relative humidity and simulated sunlight 82

4.8 Van Krevelen diagrams of halogenated SOA from α-pinene formed at 0 % relative humidity and simulated sunlight 83

4.9 Example of mass ranges and of elemental composition of SOA from catechol processed with halogens 85

4.10 Changes in the optical properties in the UV/VIS spectral range of organic aerosols due to the reaction with gaseous halogens (L: SOA formed at 0 % relative humidity und simulated sunlight; LW: SOA formed at 25 % relative humidity and simulated sunlight) 86

4.11 Changes in the differential absorbance dA due to the reaction of the organic aerosols with chlorine and bromine 87

5.1 Evolution of the mean particle diameters of the organic aerosols during processing with halogens released from a simulated salt pan. The solid line indicates the expected evolution of the particle diameter without any halogen interaction. . . 90

5.2 ATR-FTIR spectra of SOA processed with halogens released from a simulated salt pan: unprocessed SOA (solid lines), processed SOA (dashed lines). 91

5.3 UV/VIS spectra of SOA processed with halogens released from the simulated salt pan 92

5.4 Changes in the differential absorbance dA due to the reaction of the organic aerosols with halogens released from the simulated salt pan 92

List of Figures

5.5 TPP-MS spectra of the SOA processed by halogens released from the salt pan: masses 35, 37, 79 and 81, representing ^{35}Cl, ^{37}Cl, ^{79}Br, and ^{81}Br, and the calculated mass ratio to determine thermal regions were the expected isotopic ratios can be found 94

5.6 Influence of SOA from catechol on the mechanism of halogen release from the salt pan (with kind permission of N. Balzer and J. Buxmann) 95

5.7 Evolution of the mean particle diameters of the organic aerosols and the sea-salt aerosol during the halogen-SOA interaction 96

5.8 ATR-FTIR spectra of SOA processed by halogens released from the simulated sea-salt aerosol: unprocessed SOA (solid lines), processed SOA (dashed lines). . 97

5.9 Diffuse-reflectance UV/VIS spectra of the three different model aerosols after reaction with halogens from the sea-salt aerosol 97

5.10 Temperature-programmed pyrolysis mass spectra of the SOA-sea-salt interaction 98

6.1 Classification of SOA from catechol or guaiacol in the $n_C/\overline{OS_C}$-diagram 101

6.2 Halogenated SOA in different currently used diagrams 107

6.3 Simplified scheme summarizing possible effects of halogens on secondary organic aerosols: transformation of SOA to haloSOA and formation of gaseous species and secondary SOA (SSOA)................................ 109

6.4 Influence of halogens on the nucleation process and aging of SOA 110

List of Tables

2.1	Applicability of ATR crystal materials for the two electrostatic precipitators constructed	34
3.1	SOA precursors and their significant physicochemical properties	42
3.2	Assignment of infrared absorptions of SOA from α-pinene in the smog chamber	45
3.3	Assignment of infrared absorptions by SOA from α-pinene using the aerosol flow reactor	47
3.4	Assignment of ATR infrared absorptions from SOA formation from α-pinene in the smog chamber	48
3.5	Assignment of infrared absorptions from SOA from catechol and guaiacol using long-path FTIR	58
3.6	Assignment of infrared absorptions from SOA from catechol using the aerosol flow reactor	61
3.7	Assignment of infrared absorptions of SOA from catechol and guaiacol using ATR-FTIR	64
3.8	Strength of the TPP-MS signals of functional groups at their respective decomposition temperatures for the five different SOA	66
3.9	Calculated ranges of mass-spectroscopic values of SOA	70
4.1	Assignment of decrease and increase of infrared absorptions of the three different SOA reacting with halogens measured by long-path absorption infrared spectroscopy	75
4.2	Assignment of absorptions observed in the ATR spectra of the three different SOA reacting with halogens	79
4.3	Degree of halogenation of SOA formed at 0 % relative humidity and simulated sunlight	81
4.4	Main parameters of processed SOA, calculated from ICR-FT/MS spectra	84
5.1	Percentage of halogenated compounds of SOA processed with halogens released from the salt pan	94

List of Tables

5.2 Percentage of halogenated compounds of SOA processed with halogens released from sea-salt aerosol 98

6.1 Comparison of infrared spectral features of natural humic acids, HULIS, aerosol extracts and two SOA models (w–weak; m–medium; s–strong) 104

Die VDM Verlagsservicegesellschaft sucht für wissenschaftliche Verlage abgeschlossene und herausragende

Dissertationen, Habilitationen, Diplomarbeiten, Master Theses, Magisterarbeiten usw.

für die kostenlose Publikation als Fachbuch.

Sie verfügen über eine Arbeit, die hohen inhaltlichen und formalen Ansprüchen genügt, und haben Interesse an einer honorarvergüteten Publikation?

Dann senden Sie bitte erste Informationen über sich und Ihre Arbeit per Email an *info@vdm-vsg.de*.

Sie erhalten kurzfristig unser Feedback!

VDM Verlagsservicegesellschaft mbH
Dudweiler Landstr. 99   Telefon +49 681 3720 174
D - 66123 Saarbrücken   Fax   +49 681 3720 1749
www.vdm-vsg.de

Die VDM Verlagsservicegesellschaft mbH vertritt

Printed by Books on Demand GmbH, Norderstedt / Germany